TC 3-22.9

Rifle and Carbine

MAY 2016

Distribution Restriction A: Approved for public release; distribution is unlimited.

This publication supersedes FM 3-22.9, dated 12 August 2008.

Headquarters, Department of the Army

*TC 3-22.9 (FM 3-22.9)

*Training Circular
No. 3-22.9

Headquarters
Department of the Army
Washington, DC, 13 May 2016

Rifle and Carbine

Contents

		Page
	PREFACE	viii
	INTRODUCTION	ix
Chapter 1	OVERVIEW	1-1
	Safe Weapons Handling	1-2
	Rules of Firearms Safety	1-3
	Weapon Safety Status	1-4
	Weapons Control Status	1-5
	Overmatch	1-6
Chapter 2	RIFLE AND CARBINE PRINCIPLES OF OPERATION	2-1
	Army Standard Service Rifle	2-1
	Upper Receiver	2-2
	Lower Receiver	2-3
	Cycle of Function	2-4
	Cooling	2-13
Chapter 3	AIMING DEVICES	3-1
	Units of Angular Measurement	3-2
	Optics	3-8
	Thermal Sights	3-15
	Pointers / Illuminators / Lasers	3-19
Chapter 4	MOUNTABLE EQUIPMENT	4-1
	Adaptive Rail System	4-1
	Mountable Weapons	4-2
	Mountable Aiming Devices	4-4

Distribution Restriction: Approved for public release; distribution will be unlimited.

*This publication supersedes FM 3-22.9, dated 12 August 2008.

Contents

	Mountable Accessories	4-5
Chapter 5	**EMPLOYMENT**	**5-1**
	Shot Process	5-2
	Functional Elements of the Shot Process	5-3
	Target Acquisition	5-4
Chapter 6	**STABILITY**	**6-1**
	Support	6-1
	Muscle Relaxation	6-4
	Natural Point of Aim	6-4
	Recoil Management	6-5
	Shooter–Gun Angle	6-5
	Field of View	6-5
	Carry Positions	6-6
	Stabilized Firing	6-13
	Firing Positions	6-15
Chapter 7	**AIM**	**7-1**
	Common Engagements	7-1
	Common Aiming Errors	7-5
	Complex Engagements	7-6
	Target Conditions	7-10
	Environmental Conditions	7-16
	Shooter Conditions	7-22
	Compound Conditions	7-23
Chapter 8	**CONTROL**	**8-1**
	Trigger Control	8-2
	Breathing Control	8-4
	Workspace Management	8-4
	Calling the Shot	8-5
	Rate of fire	8-6
	Follow-Through	8-7
	Malfunctions	8-8
	Transition to Secondary Weapon	8-14
Chapter 9	**MOVEMENT**	**9-1**
	Movement Techniques	9-1
	Forward Movement	9-2
	Retrograde Movement	9-2
	Lateral Movement	9-3
	Turning Movement	9-4
Appendix A	**AMMUNITION**	**A-1**
Appendix B	**BALLISTICS**	**B-1**
Appendix C	**COMPLEX ENGAGEMENTS**	**C-1**

Appendix D DRILLS D-1
Appendix E ZEROING E-1
GLOSSARY Glossary-1
REFERENCES References-1
INDEX Index-1

Figures

Figure 1-1. Employment skills 1-1
Figure 1-2. Small unit range overmatch 1-8
Figure 2-1. Upper receiver 2-2
Figure 2-2. Lower receiver 2-3
Figure 2-3. Feeding example 2-5
Figure 2-4. Chambering example 2-6
Figure 2-5. Locking example 2-7
Figure 2-6. Firing example 2-8
Figure 2-7. Unlocking example 2-9
Figure 2-8. Extraction example 2-10
Figure 2-9. Ejection example 2-11
Figure 2-10. Cocking example 2-12
Figure 3-1. Minute of angle example 3-2
Figure 3-2. Mil example 3-3
Figure 3-3. Close combat optic / Rifle combat optic reticle /
    Thermal reticle examples 3-4
Figure 3-4. Stadia reticle example 3-5
Figure 3-5. Electromagnetic spectrum 3-7
Figure 3-6. Carrying handle with iron sight example 3-9
Figure 3-7. Back up iron sight 3-10
Figure 3-8. CCO Reticle, Comp M2 examples 3-12
Figure 3-9. RCO reticle example 3-14
Figure 3-10. Thermal weapon sight example 3-15
Figure 3-11. Weapon thermal sights by version 3-17
Figure 3-12. Thermal weapons sight, narrow field of view reticle
    example 3-18

Figure 3-13. Thermal weapons sight, wide field of view reticle example 3-18
Figure 3-14. AN/PEQ-2 3-21
Figure 3-15. AN/PEQ-15, ATPIAL 3-23
Figure 3-16. AN/PEQ-15A, DBAL-A2 3-25
Figure 3-17. AN/PSQ-23, STORM 3-27
Figure 4-1. M320 attached to M4 series carbine example 4-2
Figure 4-2. M203 grenade launcher example 4-2
Figure 4-3. M26 shotgun example 4-3
Figure 4-4. Bipod example 4-5
Figure 4-5. Vertical foregrip example 4-6
Figure 6-1. Stock weld 6-4
Figure 6-2. Hang carry example 6-7
Figure 6-3. Safe hang example 6-8
Figure 6-4. Collapsed low ready example 6-9
Figure 6-5. Low ready position 6-10
Figure 6-6. High ready position 6-11
Figure 6-7. Ready position or up position 6-12
Figure 6-8. Firing position stability example 6-14
Figure 6-9. Standing, unsupported example 6-16
Figure 6-10. Standing, supported example 6-17
Figure 6-11. Squatting position 6-18
Figure 6-12. Kneeling, unsupported example 6-19
Figure 6-13. Kneeling, supported example 6-20
Figure 6-14. Sitting position—crossed ankle 6-21
Figure 6-15. Sitting position—crossed-leg 6-22
Figure 6-16. Sitting position—open leg 6-23
Figure 6-17. Prone, unsupported example 6-24
Figure 6-18. Prone, supported example 6-25
Figure 6-19. Prone, roll-over example 6-26
Figure 6-20. Reverse roll-over prone firing position 6-27
Figure 7-1. Horizontal weapon orientation example 7-2
Figure 7-2. Vertical weapons orientation example 7-3
Figure 7-3. Front sight post/reticle aim focus 7-4

Contents

Figure 7-4. Immediate hold locations for windage and lead example 7-8
Figure 7-5. Immediate hold locations for elevation (range) example 7-9
Figure 7-6. Front sight post method example 7-11
Figure 7-7. Immediate holds for range to target 7-13
Figure 7-8. Immediate holds for moving targets example 7-14
Figure 7-9. Oblique target example 7-15
Figure 7-10. Wind value 7-17
Figure 7-11. Wind effects 7-18
Figure 7-12. Wind hold example 7-20
Figure 7-13. Compound wind and lead determination example 7-23
Figure 8-1. Arc of movement example 8-2
Figure 8-2. Natural trigger finger placement 8-3
Figure 8-3. Workspace example 8-5
Figure 8-4. Malfunction corrective action flow chart 8-12
Figure A-1. Small arms ammunition cartridges A-1
Figure A-2. Cartridge case A-2
Figure A-3. Propellant A-3
Figure A-4. 5.56mm primer detail A-4
Figure A-5. Bullet example, Armor-piercing cartridge A-5
Figure A-6. Ball cartridge A-6
Figure A-7. Ball with tracer cartridge A-6
Figure A-8. Armor-piercing cartridge A-7
Figure A-9. Short range training ammunition cartridge A-7
Figure A-10. Blank cartridge A-8
Figure A-11. Close combat mission capability kit cartridge A-8
Figure A-12. Dummy cartridge A-9
Figure B-1. Internal ballistic terms B-2
Figure B-2. External ballistic terms B-3
Figure B-3. Trajectory B-4
Figure B-4. Lethal zone example B-11
Figure C-1. Mil Relation Formula example C-2
Figure C-2. Standard dismount threat dimensions example C-3
Figure C-3. RCO range determination using the bullet drop compensator reticle C-4

Figure C-4. Reticle relationship using a stadiametric reticle example ... C-5
Figure C-5. Deliberate lead formula example ... C-7
Figure C-6. Deliberate trapping method example ... C-8
Figure C-7. Oblique target example ... C-9
Figure C-8. Wind value ... C-11
Figure C-9. Wind effects ... C-12
Figure C-10. Wind formula and ballistics chart example ... C-14
Figure C-11. Hold-off example ... C-15
Figure C-12. Quick high angle formula example ... C-17
Figure C-13. Compound wind and lead determination example ... C-18
Figure E-1. Wind effects on zero at 300 meters ... E-3
Figure E-2. M16A2 / M16A4 weapons 25m zero target ... E-5
Figure E-3. M4-/M16-series weapons 25m zero short range and pistol marksmanship target ... E-6
Figure E-4. Grouping ... E-7
Figure E-5. Marking shot groups ... E-8
Figure E-6a. Horizontal diagnostic shots ... E-12
Figure E-6b. Vertical diagnostic shots ... E-13

Tables

Table 1-1. Weapons Safety Status for M4- and M16-Series Weapons ... 1-5
Table 1-2. Weapons Control Status ... 1-5
Table 2-1. Model Version Firing Methods Comparison ... 2-4
Table 3-1. Laser Aiming Devices for the M4 and M16 ... 3-19
Table 4-1. Attachment Related Technical Manuals and Mounting ... 4-4
Table 5-1. Shot Process example ... 5-2
Table A-1. Small Arms Color Coding and Packaging Markings ... A-10
Table A-2. 5-56mm, M855, Ball ... A-11
Table A-3. 5.56mm, M855A1, Enhanced Performance Round (EPR), Ball ... A-12
Table A-4. 5.56mm, M856A1, Tracer ... A-13
Table A-5. 5.56mm, Mk301, MOD 0, DIM Tracer ... A-14

Table A-6. 5.56mm, M995, Armor Piercing A-15
Table A-7. 5.56mm, M862, Short Range Training Ammunition A-16
Table A-8. 5.56mm, M1037, Short Range Training Ammunition A-17
Table A-9. 5.56mm, M1042 Close Combat Mission Capability Kit A-18
Table A-10. 5.56mm, M200, Blank A-19
Table C-1. Standard holds beyond zero distance example C-16

Preface

Training Circular (TC) 3-22.9 provides Soldiers with the critical information for their rifle or carbine and how it functions, its capabilities, the capabilities of the optics and ammunition, and the application of the functional elements of the shot process.

TC 3-22.9 uses joint terms where applicable. Selected joint and Army terms and definitions appear in both the glossary and the text. Terms for which TC 3-22.9 is the proponent publication (the authority) are italicized in the text and are marked with an asterisk (*) in the glossary. Terms and definitions for which TC 3-22.9 is the proponent publication are boldfaced in the text. For other definitions shown in the text, the term is italicized and the number of the proponent publication follows the definition.

The principal audience for TC 3-22.9 is all members of the profession of arms. Commanders and staffs of Army headquarters serving as joint task force or multinational headquarters should also refer to applicable joint or multinational doctrine concerning the range of military operations and joint or multinational forces. Trainers and educators throughout the Army will also use this publication.

Commanders, staffs, and subordinates ensure that their decisions and actions comply with applicable United States, international, and in some cases host-nation laws and regulations. Commanders at all levels ensure that their Soldiers operate in accordance with the law of war and the rules of engagement. (See FM 27-10.)

This publication applies to the active Army, the Army National Guard (ARNG)/Army National Guard of the United States (ARNGUS), and the United States Army Reserve (USAR). Unless otherwise stated in this publication, masculine nouns and pronouns do not refer exclusively to men.

Uniforms depicted in this manual were drawn without camouflage for clarity of the illustration.

The proponent of this publication is United States (U.S.) Army Maneuver Center of Excellence (MCoE). The preparing agency is the MCoE, Fort Benning, Georgia. You may submit comments and recommended changes in any of several ways—U.S. mail, e-mail, fax, or telephone—as long as you use or follow the format of DA Form 2028, (*Recommended Changes to Publications and Blank Forms*). Contact information is as follows:

E-mail:	usarmy.benning.mcoe.mbx.doctrine@mail.mil
Phone:	COM 706-545-7114 or DSN 835-7114
Fax:	COM 706-545-8511 or DSN 835-8511
U.S. mail:	Commander, MCoE
	Directorate of Training and Doctrine (DOTD)
	Doctrine and Collective Training Division
	ATTN: ATZB-TDD
	Fort Benning, GA 31905-5410

Introduction

This manual is comprised of nine chapters and five appendices, and is specifically tailored to the individual Soldier's use of the M4- or M16-series weapon. This TC provides specific information about the weapon, aiming devices, attachments, followed by sequential chapters on the tactical employment of the weapon system.

The training circular itself is purposely organized in a progressive manner, each chapter or appendix building on the information from the previous section. This organization provides a logical sequence of information which directly supports the Army's training strategy for the weapon at the individual level.

Chapters 1 through 4 describe the weapon, aiming devices, mountable weapons, and accessories associated with the rifle and carbine. General information is provided in the chapters of the manual, with more advanced information placed in appendix A, Ammunition, and appendix B, Ballistics.

Chapters 5 through 9 provide the employment, stability, aiming, control and movement information. This portion focuses on the Solider skills needed to produce well aimed shots. Advanced engagement concepts are provided in appendix C of this publication. Appendix D of this publication provides common tactical drills that are used in training and combat that directly support tactical engagements. Finally, appendix E of this publication, is provided at a common location in this and future weapons publications to provide a common location for reference.

This manual does not cover the specific rifle or carbine training strategy, ammunition requirements for the training strategy, or range operations. These areas will be covered in separate training circulars.

Conclusion

TC 3-22.9 applies to all Soldiers, regardless of experience or position. This publication is designed specifically for the Soldier's use on the range during training, and as a reference while deployed.

This page intentionally left blank.

Chapter 1

Overview

This TC is designed to provide Soldiers the critical information on their rifle or carbine to properly and effectively engage and destroy threats in a direct fire engagement. It relies on the Soldier's understanding of the weapon, how it functions, its capabilities, the capabilities of the optics and ammunition, and how to properly employ those capabilities to achieve mastery through the application of the functional elements of the shot process.

This chapter describes the principles of proper weapons handling, tactical applications and control measures for handling the weapons, and an overview of the concepts of overmatch as it pertains to a Soldier's individual weapon.

1-1. Each Soldier is responsible for placing accurate and effective fires on threat targets with their individual weapon. This manual defines the functional elements of the shot process, the principles of operation of the weapon, the characteristics and description of ballistics and ammunition, and the various engagement techniques that are essential to build Soldier proficiency with their weapon. It includes standard drills and techniques that assist the Soldier to build, improve, and sustain their skills to achieve accurate and precise shots consistently during combat operations (see figure 1-1).

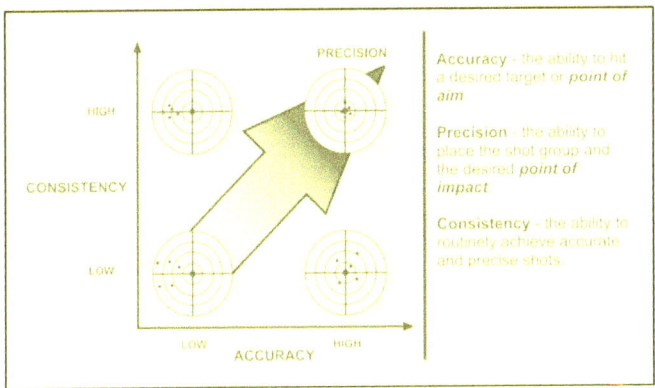

Figure 1-1. Employment skills

SAFE WEAPONS HANDLING

1-2. Safe weapons handling procedures are a consistent and standardized way for Soldiers to handle, operate, and employ the weapon safely and effectively. Weapons handling is built on three components; the Soldier, the weapon, and the environment:

- The **Soldier** must maintain situational understanding of friendly forces, the status of the weapon, and the ability to evaluate the environment to properly handle any weapon. The smart, adaptive, and disciplined Soldier is the primary safety mechanism for all weapons under his control.
- The **weapon** is the primary tool of the Soldier to defeat threats in combat. The Soldier must know of and how to operate the mechanical safeties built into the weapons they employ, as well as the principles of operation for those weapons.
- The **environment** is the Soldier's surroundings. The Soldier must be aware of muzzle discipline, the nature of the target, and what is behind it.

1-3. To safely and effectively handle weapons, Soldiers must be cognitively aware of three distinct weapons handling measures:

- **The rules of firearms safety.**
- **Weapons safety status.**
- **Weapons control status.**

1-4. These measures directly support the components of safe weapons handling. They are designed to provide redundant safety measures when handling any weapon or weapon system, not just rifles and carbines.

1-5. This redundancy allows for multiple fail-safe measures to provide the maximum level of safety in both training and operational environments. A Soldier would have to violate two of the rules of firearms safety or violate a weapon safety status in order to have a negligent discharge.

Note. Unit standard operating procedures (SOPs), range SOPs, or the operational environment may dictate additional safety protocols; however, the rules of firearms safety are always applied. If a unit requires Soldiers to violate these safety rules for any reason, such as for the use of blank rounds or other similar training munitions during training, the unit commander must take appropriate risk mitigation actions.

Overview

RULES OF FIREARMS SAFETY

1-6. The Rules of Firearms Safety are standardized for any weapon a Soldier may employ. Soldiers must adhere to these precepts during training and combat operations, regardless of the type of ammunition employed, except as noted above.

Rule 1: Treat Every Weapon as if it is Loaded

1-7. Any weapon handled by a Soldier must be treated as if it is loaded and prepared to fire. Whether or not a weapon is loaded should not affect how a Soldier handles the weapon in any instance.

1-8. Soldiers must take the appropriate actions to ensure the proper weapon status is applied during operations, whether in combat or training.

Rule 2: Never Point the Weapon at Anything You Do Not Intend to Destroy

1-9. Soldiers must be aware of the orientation of their weapon's muzzle and what is in the path of the projectile if the weapon fires. Soldiers must ensure the path between the muzzle and target is clear of friendly forces, noncombatants, or anything the Soldier does not want to strike.

1-10. When this is unavoidable, the Soldier must minimize the amount of time the muzzle is oriented toward people or objects they do not intend to shoot, while simultaneously applying the other three rules of fire arms safety.

Rule 3: Keep Finger Straight and Off the Trigger Until Ready to Fire

1-11. Soldiers must not place their finger on the trigger unless they intend to fire the weapon. The Soldier is the most important safety feature on any weapon. Mechanical safety devices are not available on all types of weapons. When mechanical safeties are present, Soldiers must not solely rely upon them for safe operation knowing that mechanical measures may fail.

1-12. Whenever possible, Soldiers should move the weapon to mechanical safe when a target is not present. If the weapon does not have a traditional mechanical safe, the trigger finger acts as the primary safety.

Rule 4: Ensure Positive Identification of the Target and its Surroundings

1-13. The disciplined Soldier can positively identify the target and knows what is in front of and what is beyond it. The Soldier is responsible for all bullets fired from their weapon, including the projectile's final destination.

1-14. Application of this rule minimizes the possibility of fratricide, collateral damage, or damage to infrastructure or equipment. It also prepares the Soldier for any follow-on shots that may be required.

WEAPON SAFETY STATUS

1-15. The readiness of a Soldier's weapon is termed as its *weapon safety status (WSS)*. It is a standard code that uses common colors (green, amber, and red) to represent the level of readiness for a given weapon.

1-16. Each color represents a specific series of actions that are applied to a weapon. They are used in training and combat to place or maintain a level of safety relevant to the current task or action of a Soldier, small unit, or group.

Note. If the component, assembly, or part described is unclear, refer to the weapon's technical manual (TM) or chapter 2 of this publication.

1-17. The following WSS are used for all M4- and M16-series weapons:

GREEN

1-18. The weapon's magazine is removed, its chamber is empty, its bolt is locked open or forward, and the selector is set to SAFE.

Note. The command given to direct a GREEN safety status is GREEN AND CLEAR or GO GREEN.

AMBER

1-19. A magazine is locked into the magazine well of the weapon, the bolt is forward on an EMPTY chamber, the ejection port cover should be CLOSED, and the selector should be set to SAFE.

Note. The command given to direct an AMBER safety status is GO AMBER.

RED

1-20. The weapon's magazine is inserted, a round is in the chamber, the bolt is forward and locked, the ejection port cover is closed, and the selector is set to SAFE.

Note. Following Rule 4, the Soldier places the weapon's selector to SEMI, BURST, or AUTO as appropriate for the weapon, and places their finger on the trigger *only* when ready to engage.

1-21. Table 1-1, on page 1-5, shows the WSS for the M4- and M16-series weapons.

Overview

Table 1-1. Weapons Safety Status for M4- and M16-Series Weapons

STATUS	GREEN	AMBER	RED
FUNCTION	CLEAR	PREPARED	READY, SAFE
COMMANDS:	GREEN AND CLEAR	LOAD MAGAZINE	MAKE READY
Ammunition	None	Magazine in	Magazine in/ round chambered
Bolt	Open or forward	Forward	Forward
Chamber	Empty	Empty	Locked
Safety	Safe	Safe	Safe
Trigger	Off	Off	Off

WEAPONS CONTROL STATUS

1-22. A *weapons control status (WCS)* is a tactical method of fire control given by a leader that incorporates the tactical situation, rules of engagement for the area of operations, and expected or anticipated enemy contact. The WCS outlines the target identification conditions under which friendly elements may engage a perceived threat with direct fire.

1-23. Table 1-2 provides a description of the standard WCS used during tactical operations, both in training and combat. They describe when the firer is authorized to engage a threat target once the threat conditions have been met.

Table 1-2. Weapons Control Status

WEAPONS CONTROL STATUS	DESCRIPTION
WEAPONS HOLD	Engage only if engaged or ordered to engage.
WEAPONS TIGHT	Engage only if target *is positively identified* as *enemy*.
WEAPONS FREE	Engage targets *not positively identified* as *friendly*.

1-24. A weapon control status and a weapons safety status are both implemented and available to leaders to prevent fratricide and limit collateral damage. These postures or statuses are typically suited to the area of operation or type of mission and should always be clearly outlined to all Soldiers, typically in the operations order (OPORD), warning order (WARNORD), or fragmentary order (FRAGORD).

Chapter 1

OVERMATCH

1-25. Overmatch is the Soldier applying their learned skills, employing their equipment, leveraging technology, and applying the proper force to create an unfair fight in favor of the Soldier. To achieve and maintain overmatch against any threat, this publication focuses on providing information that develops the Soldier's direct fire engagement skills using the following attributes:

- **Smart** – the ability to routinely generate understanding through changing conditions.
- **Fast** – the ability to physically and cognitively outmaneuver adversaries.
- **Lethal** – deadly in the application of force.
- **Precise** – consistently accurate in the application of power to ensure deliver of the right effects in time, space, and purpose.

1-26. This requires the Soldier to understand the key elements that build the unfair advantage and exploit them at every opportunity during tactical operations. The components of overmatch are:

- **Target detection, acquisition, and identification** – the ability of the Soldier to detect and positively identify any suspected target as hostile at greater distances than their adversary. This relies upon Soldier training and their ability to leverage the capabilities of their optics, thermals, and sensors.
- **Engagement range** – provide the Soldier with weapons, aiming devices, and ammunition capable of striking and defeating a threat at a greater range than the adversary can detect or engage the friendly force with effective fires.
- **Limited visibility** – provide the Soldier to make operations during limited visibility an advantage through technology and techniques, and compound their adversary's disadvantages during those conditions.
- **Precision** – provide a weapon and ammunition package that enhances the Soldier's consistent application of shots with a level of precision greater than the adversary's.
- **Speed** – the weapon, aiming devices, and accessories a Soldier employs must seamlessly work in unison, be intuitive to use, and leverage natural motion and manipulations to facilitate rapid initial and subsequent shots during an engagement at close quarters, mid-, and extended ranges.
- **Terminal performance** – ensures that precise shots delivered at extended ranges provide the highest probability to defeat the threat through exceptional ballistic performance.

1-27. Although not a component of overmatch, exceptional training is critical to create smart, fast, lethal, and precise Soldiers. Training builds proficiency in a progressive, logical, and structured manner and provides Soldiers the skills necessary to achieve overmatch against any adversary. This requires the training program to provide experience to the Soldier in all the components of overmatch to their fullest extent possible in the shortest amount of time.

Overview

TARGET DETECTION, ACQUISITION, AND IDENTIFICATION

1-28. The first component of overmatch at the Soldier level is the ability to detect targets as far away as possible during limited and low visibility conditions. This manual describes the aiming devices for the service rifle that enhance the Soldier's target detection and acquisition skills. The Soldier must be able to detect, acquire, and identify targets *at ranges beyond* the maximum effective range of their weapon and ammunition.

1-29. This publication also provides key recognition information to build the Soldier's skills in correctly identifying potential targets as friend, foe, or noncombatant (neutral) once detected.

ENGAGEMENT RANGE

1-30. To ensure small unit success, the Soldier requires weapon systems that can effectively engage threats at ranges greater than those of their adversaries. This creates a standoff distance advantage that allows friendly forces to destroy the target outside the threat's maximum effective range.

1-31. Range overmatch provides a tactical engagement buffer that accommodates the Soldier's time to engage with precision fires. For example, a Soldier that has the capability to effectively engage personnel targets at a range of 500 meters will have range overmatch of 10 to 20 percent over a threat rifleman. That 10 to 20 percent range difference is equivalent to a distance of 40 to 80 meters, which is approximately the distance a maneuvering threat can traverse in 15 to 40 seconds.

1-32. Figure 1-2 portrays the battlefield from the Soldier's perspective. With mobile, maneuvering threats, the target acquisition capabilities must compliment the engagement of those threats at the maximum effective range of the weapon, optic, and ammunition.

Chapter 1

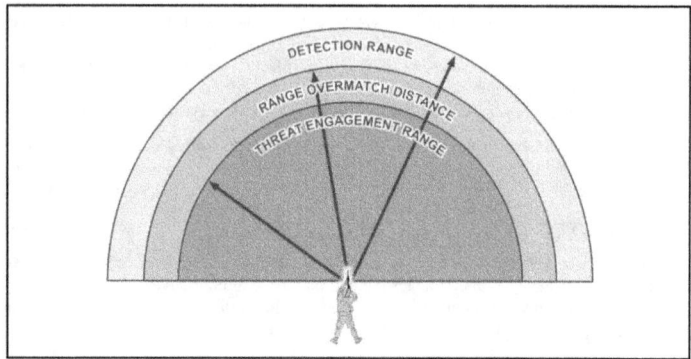

Figure 1-2. Small unit range overmatch

LIMITED VISIBILITY

1-33. Soldiers must be able to detect, acquire, identify, and engage threats in all light conditions, regardless of the tactical situation. To provide that capability, aiming devices are provided that minimize the effects of limited visibility, but not completely.

1-34. Image intensifiers and thermal optics provide a significant overmatch capability, but they also have limitations and disadvantages. A general discussion of their capabilities, particularly what those systems can view within the spectrum of light is provided. Soldiers must understand what can be "seen" or viewed and what cannot when using their assigned equipment. Understanding the advantages and limitations of their equipment has a direct impact on force protection, fratricide and collateral damage prevention, and maintaining overmatch during tactical operations.

PRECISION

1-35. The Army standard service rifle is designed with a specific level of accuracy out to its maximum effective range. This level of accuracy is more consistent and reliable through the use of magnified aiming devices and superior ammunition. The Soldier must build the skills to use them effectively to deliver precision fires during tactical engagements.

SPEED

1-36. The close fight requires rapid manipulations, a balance of speed and accuracy, and very little environmental concerns. Soldiers must move quickly and efficiently through their manipulations of the fire control to maintain the maximum amount of muzzle orientation on the threat through the shot process. This second-nature efficiency of movement only comes from regular practice, drills, and repetition.

1-37. The foundation of speed of action is built through understanding the weapon, ammunition, ballistics, and principles of operation of the associated aiming devices. It is reinforced during drills (appendix D), and the training program of the unit.

1-38. The goal of training to overmatch is to increase the speed at which the Soldier detects a threat, identifies it as hostile, and executes the shot process with the desired target effect. This manual is constructed to provide the requisite information in a progressive manner to build and reinforce Soldier understanding, confidence, and ability to execute tactical operations with speed and smooth fluidity of motion.

TERMINAL BALLISTIC PERFORMANCE

1-39. Terminal ballistic performance is the actions of a projectile from the time it strikes an object downrange until it comes to rest. The ammunition used with the service rifle performs exceptionally well out to its maximum effective range and beyond. This manual provides information on the various munition types available for training and combat, their capabilities and purpose, and the service (combat) round's terminal ballistic performance (see appendix A, Ammunition, and appendix B, Ballistics).

1-40. Soldiers must understand the capabilities of their ammunition, whether designed for training or combat use. That understanding creates a respect for the weapon and ammunition, reinforces the precepts of safe weapons handling, and an understanding of the appropriate skills necessary to deliver lethal fires.

1-41. Soldiers that understand the "how" and "why" of their weapon system, aiming devices, ammunition, and procedures work or function develops a more comprehensive understanding. That level of understanding, coupled with a rigorous training program that builds and strengthens their skills create more proficient Soldiers. The proficiencies and skills displayed during training translate into smart, fast, lethal and precise Soldiers for the small unit during decisive action combat operations.

This page intentionally left blank.

Chapter 2

Rifle and Carbine Principles of Operation

This chapter provides the general characteristics, description, available components, and principles of operation for the M4- and M16-series weapons. It provides a general overview of the mechanics and theory of how weapons operate, key terms and definitions related to their functioning, and the physical relationship between the Soldier, the weapon, and the optics/equipment attached to the weapon.

ARMY STANDARD SERVICE RIFLE

2-1. The Army standard service rifle is either the M16-series rifle or M4-series carbine. These weapons are described as a lightweight, 5.56-mm, magazine-fed, gas-operated, air-cooled, shoulder-fired rifle or carbine. They fire in semiautomatic (single-shot), three-round burst, or in automatic mode using a selector lever, depending on the variant. The weapon system has a standardized mounting surface for various optics, pointers, illuminators, and equipment, to secure those items with common mounting and adjustment hardware.

2-2. Each service rifle weapon system consists of components, assemblies, subassemblies, and individual parts. Soldiers must be familiar with these items and how they interact during operation.

- **Components** are uniquely identifiable group of fitted parts, pieces, assemblies or subassemblies that are required and necessary to perform a distinctive function in the operation of the weapon. Components are usually removable in one piece and are considered indivisible for a particular purpose or use.
- **Assemblies** are a group of subassemblies and parts that are fitted to perform specific set of functions during operation, and cannot be used independently for any other purpose.
- **Subassemblies** are a group of parts that are fitted to perform a specific set of functions during operation. Subassemblies are compartmentalized to complete a single specific task. They may be grouped with other assemblies, subassemblies and parts to create a component.
- **Parts** are the individual items that perform a function when attached to a subassembly, assembly, or component that serves a specific purpose.

2-3. Each weapon consists of two major components: the upper receiver and the lower receiver. These components are described below including their associated assemblies, subassemblies, and parts.

Chapter 2

UPPER RECEIVER

2-4. An aluminum receiver helps reduce the overall weight of the rifle/carbine and allows for mounting of equipment and accessories. The upper receiver consists of the following (see figure 2-1):

- Barrel assembly.
  - **Barrel**. The bore and chamber of the barrel are chrome-plated to reduce wear and fouling over the life of the weapon.
  - **Flash hider or compensator**. Located at the end of the barrel, is provided to reduce the signature of the weapon during firing and reduce barrel movement off target during firing.
  - **Sling swivel**. The attachment hardware for the sling system used to properly carry the weapon.
  - **Front sight assembly**. Includes an adjustable front sight post that facilitates zeroing the weapon, serves as the forward portion of the iron sight or back up iron sight, and assists with range determination.
  - **Adapter rail system (ARS)**. Provided in varying lengths, depending on the variant applied. Used to attach common aiming devices or accessories.
  - **Slip ring**. Provides a spring loaded locking mechanism for the weapon's hand guards.
  - **Ejection port**. Provides an opening in the upper receiver to allow ammunition or spent casing ejection from the weapon.
  - **Ejection port cover**. Provides a dust cover for the ejection port, protecting the upper receiver and bolt assembly from foreign objects.
  - **Forward assist assembly**. Provides a Soldier applied mechanical assist to the bolt assembly during operations.

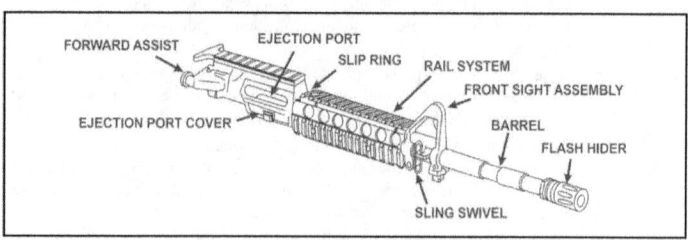

Figure 2-1. Upper receiver

Rifle and Carbine Principles of Operation

LOWER RECEIVER

2-5. The lower receiver shown in figure 2-2, on page 2-3, consists of the following components, assemblies, and parts:

- **Trigger assembly.** Provides the trigger, pins, springs, and other mechanical components necessary to fire the weapon.
- **Bolt catch.** A mechanical lever that can be applied to lock the bolt to the rear by the Soldier, or automatically during the cycle of function when the magazine is empty (see page 2-4).
- **Rifle grip.** An ambidextrous pistol-type handle that assists in recoil absorption during firing.
- **Magazine catch assembly.** Provides a simple, spring-loaded locking mechanism to secure the magazine within the magazine well. Provides the operator an easy to manipulate, push-to-release textured button to release the magazine from the magazine well during operation.
- **Buttstock assembly.** Contains the components necessary for proper shoulder placement of the weapon during all firing positions, returning the bolt assembly to battery, and managing the forces of recoil during operation.
  - The M4/M4A1-series carbine has a four position collapsible buttstock assembly: Closed, ½ open, ¾ open, and fully-open.
  - M16-series rifles have a fixed buttstock with cleaning kit compartment or an applied modified work order (MWO) collapsible buttstock.
- **Action spring.** Provides the stored energy to return the bolt carrier assembly back into battery during operation.
- **Lower receiver extension.** Provides space for the action spring and buffer assembly during operation.

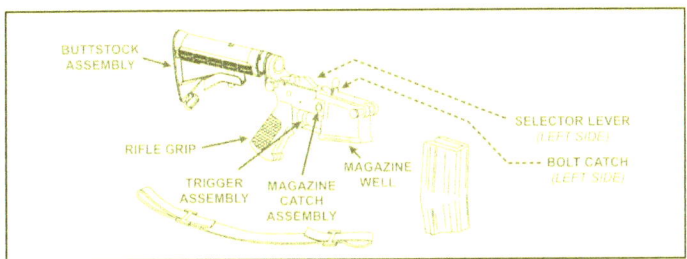

Figure 2-2. Lower receiver

2-6. Additional information on the characteristics and components of the M4/M4A1/M16-series weapons can be found in technical manual (TM) 9-1005-319-10. Soldiers will use the technical manual for preventative maintenance checks and services (PMCS).

and operation under normal conditions, as well as more detailed information on the principles of operation.

2-7. Each variant of the rifle and carbine have subtle capabilities differences. The primary differences are shown in table 2-1, and are specific to the weapon's selector switch, buttstock, and barrel length.

Table 2-1. Model Version Firing Methods Comparison

Weapon	Selector Switch Position			Buttstock	Barrel Length
M16A2	SAFE	SEMI	BURST	Full	20 inches
M16A3	SAFE	SEMI	AUTO	Full	20 inches
M16A4	SAFE	SEMI	BURST	Full	20 inches
M4	SAFE	SEMI	BURST	Collapsible	14.5 inches
M4A1	SAFE	SEMI	AUTO	Collapsible	14.5 inches
Legend: SEMI: semi-automatic firing selection AUTO: fully automatic firing selection BURST: three-round burst firing selection					

CYCLE OF FUNCTION

2-8. The *cycle of function* is the mechanical process a weapon follows during operation. The information provided below is specific to the cycle of function as it pertains specifically to the M4- and M16-series weapons.

2-9. The cycle starts when the rifle is ready with the bolt locked to the rear, the chamber is clear, and a magazine inserted into the magazine well with at least one cartridge. From this state, the cycle executes the sequential phases of the cycle of functioning to fire a round and prepare the weapon for the next round. The phases of the cycle of function in order are—

- Feeding.
- Chambering.
- Locking.
- Firing.
- Unlocking.
- Extracting.
- Ejecting.
- Cocking.

2-10. For the weapon to operate correctly, semiautomatic and automatic weapons require a *system of operation* to complete the cycle of functioning. The M4- and M16-series weapons use a direct impingement gas operating system. This system uses a portion of the high pressure gas from the cartridge being fired to physically move the assemblies and subassemblies in order to complete the cycle of function.

Rifle and Carbine Principles of Operation

FEEDING

2-11. Feeding is the process of mechanically providing a cartridge of ammunition to the entrance of the chamber (see figure 2-3).

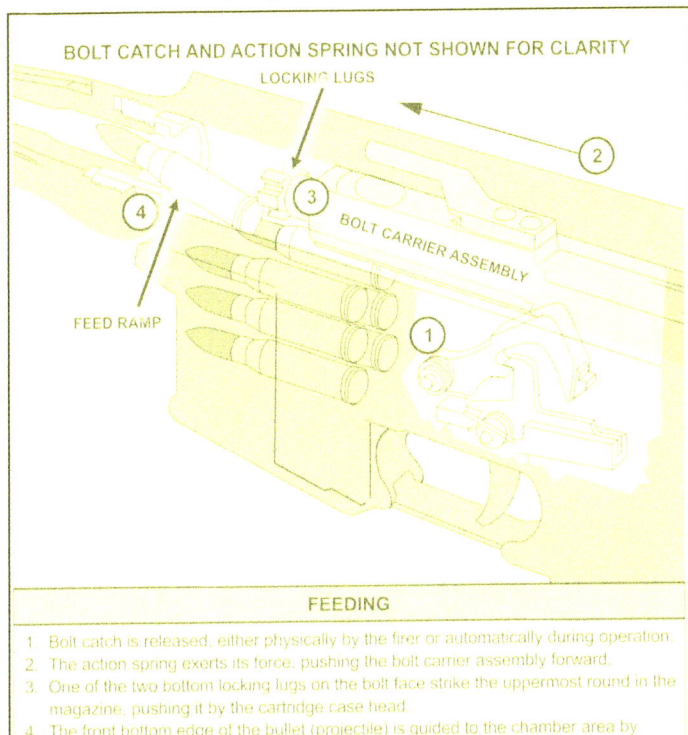

Figure 2-3. Feeding example

Chapter 2

CHAMBERING

2-12. Chambering is the continuing action of the feeding round into the chamber of the weapon (see figure 2-4).

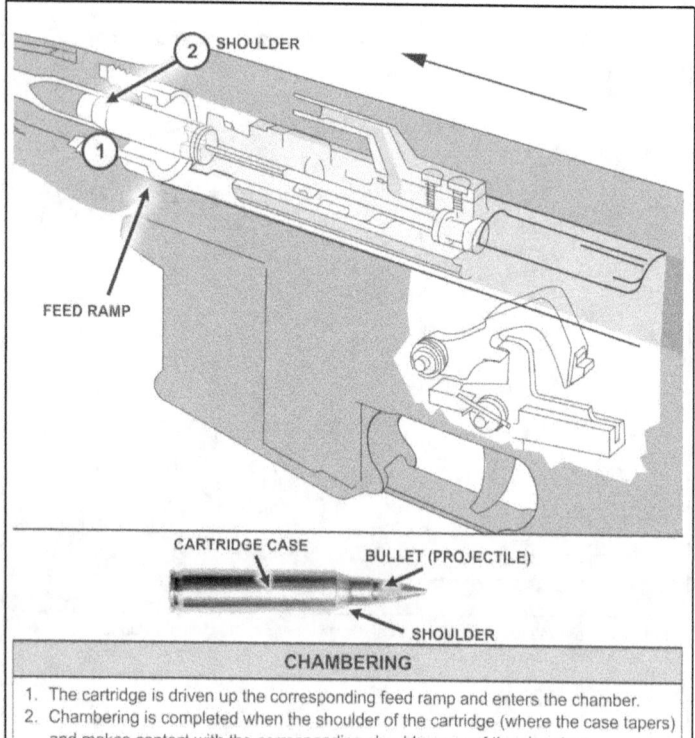

Figure 2-4. Chambering example

Rifle and Carbine Principles of Operation

LOCKING

2-13. Locking is the process of creating a mechanical grip between the bolt assembly and chamber with the appropriate amount of headspace (clearance) for safe firing (see figure 2-5). With the M4- and M16-series weapons, locking takes place simultaneously with the final actions of chambering.

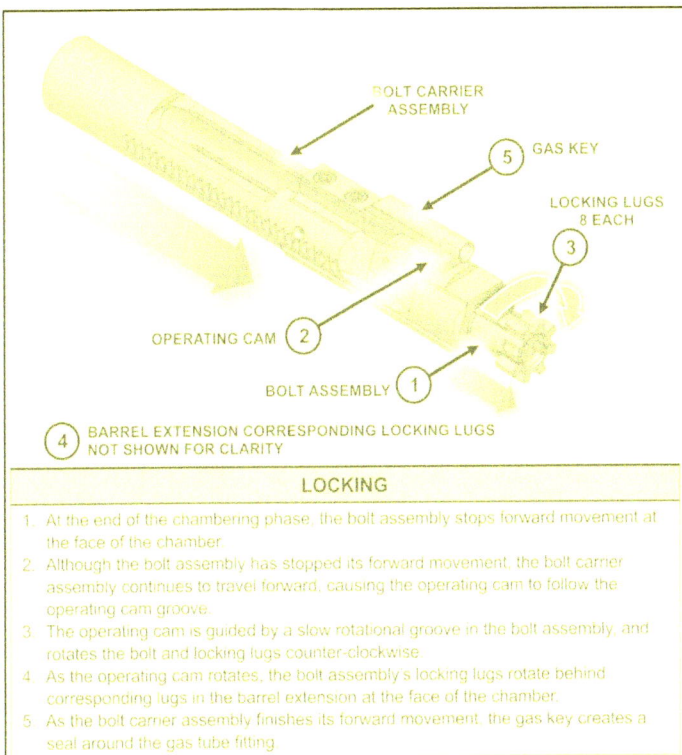

LOCKING

1. At the end of the chambering phase, the bolt assembly stops forward movement at the face of the chamber.
2. Although the bolt assembly has stopped its forward movement, the bolt carrier assembly continues to travel forward, causing the operating cam to follow the operating cam groove.
3. The operating cam is guided by a slow rotational groove in the bolt assembly, and rotates the bolt and locking lugs counter-clockwise.
4. As the operating cam rotates, the bolt assembly's locking lugs rotate behind corresponding lugs in the barrel extension at the face of the chamber.
5. As the bolt carrier assembly finishes its forward movement, the gas key creates a seal around the gas tube fitting.

Figure 2-5. Locking example

Chapter 2

FIRING

2-14. Firing is the finite process of initiating the primer detonation of the cartridge and continues through shot-exit of the projectile from the muzzle (see figure 2-6).

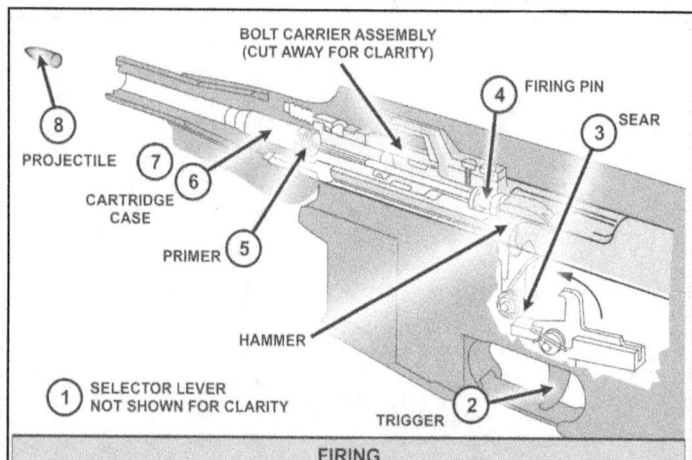

FIRING

1. Firing is initiated when the weapon is on SEMI or BURST/AUTO, a round is chambered, and the trigger is pressed.
2. Once the trigger is pressed, the sear releases the hammer.
3. As the sear releases the hammer, the hammer spring exerts its force rotating the hammer forward through the bolt carrier assembly opening.
4. The hammer strikes the firing pin, pushing the loaded firing pin forward to the primer of the cartridge case.
5. The firing pin crushes the cartridge primer, igniting the primer's charge.
6. The burning primer charge ignites the round's propellant.
7. As the propellant burns, the cartridge case expands to the fullest extent of the chamber area and to the face of the bolt, sealing the gases within the bore.
8. The expending gas propels the projectile down the length of the bore.

Figure 2-6. Firing example

Rifle and Carbine Principles of Operation

UNLOCKING

2-15. Unlocking is the process of releasing the locking lugs on the bolt face from the corresponding recesses on the barrel extension surrounding the chamber area (see figure 2-7).

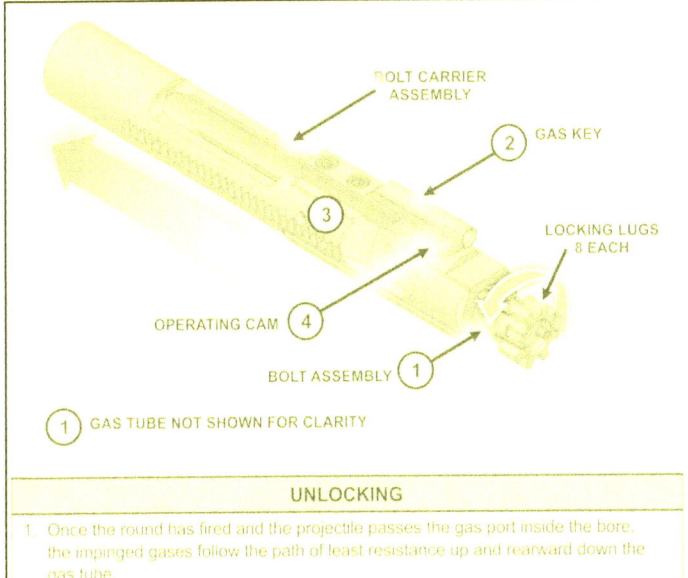

UNLOCKING

1. Once the round has fired and the projectile passes the gas port inside the bore, the impinged gases follow the path of least resistance up and rearward down the gas tube.
2. The rearward gases traveling down the gas tube apply their force through the gas key, filling the area inside the bolt carrier assembly behind the gas rings on the bolt assembly.
3. The expanding gas pushes the bolt carrier assembly rearward while the gas seal rings retain the bolt forward.
4. While the bolt carrier assembly moves rearward, the operating cam follows the operating cam groove, and rotates the bolt and its locking lugs clockwise.

Figure 2-7. Unlocking example

Chapter 2

EXTRACTING

2-16. Extracting is the removal of the expended cartridge case from the chamber by means of the extractor (see figure 2-8).

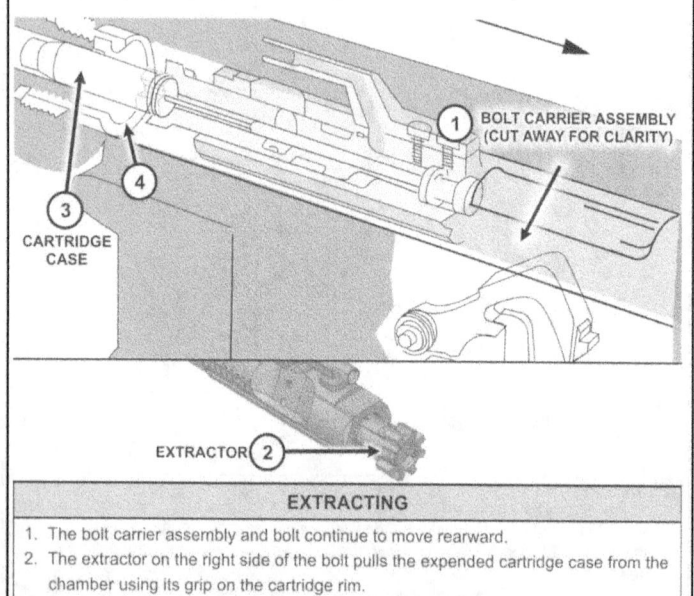

EXTRACTING

1. The bolt carrier assembly and bolt continue to move rearward.
2. The extractor on the right side of the bolt pulls the expended cartridge case from the chamber using its grip on the cartridge rim.
3. Initially, the extractor breaks the seal of the expended cartridge case from the chamber area.
   The extractor's spring loaded force maintains pressure on the cartridge rim while it continues to pull the cartridge out of the chamber.
4. The extracting phase continues until the cartridge case is clear of the chamber area but has not exited the weapon.

Figure 2-8. Extraction example

Rifle and Carbine Principles of Operation

EJECTING

2-17. Ejecting is the removal of the spent cartridge case from the weapon itself (see figure 2-9.)

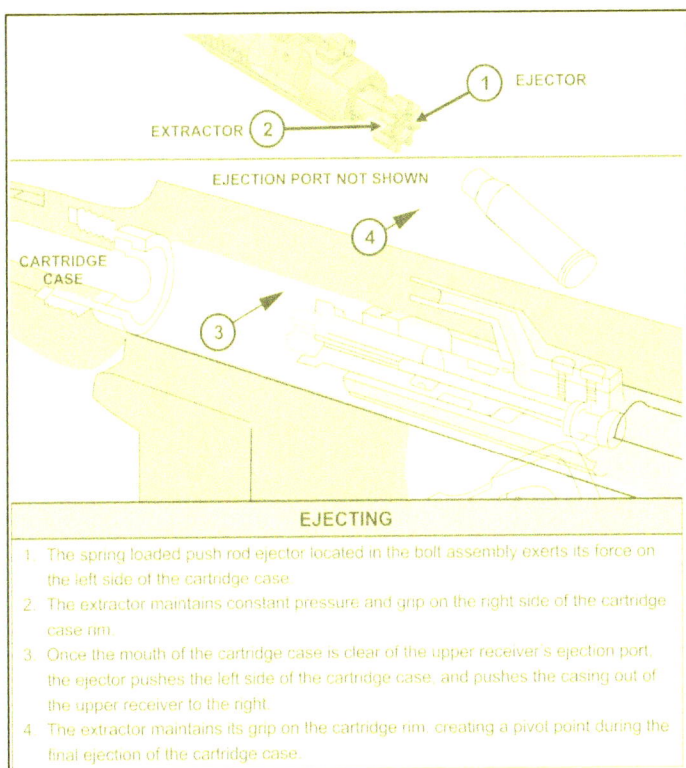

EJECTING

1. The spring loaded push rod ejector located in the bolt assembly exerts its force on the left side of the cartridge case.
2. The extractor maintains constant pressure and grip on the right side of the cartridge case rim.
3. Once the mouth of the cartridge case is clear of the upper receiver's ejection port, the ejector pushes the left side of the cartridge case, and pushes the casing out of the upper receiver to the right.
4. The extractor maintains its grip on the cartridge rim, creating a pivot point during the final ejection of the cartridge case.

Figure 2-9. Ejection example

Chapter 2

COCKING

2-18. Cocking is the process of mechanically positioning the trigger assembly's parts for firing (see figure 2-10). The cocking phase completes the full cycle of functioning.

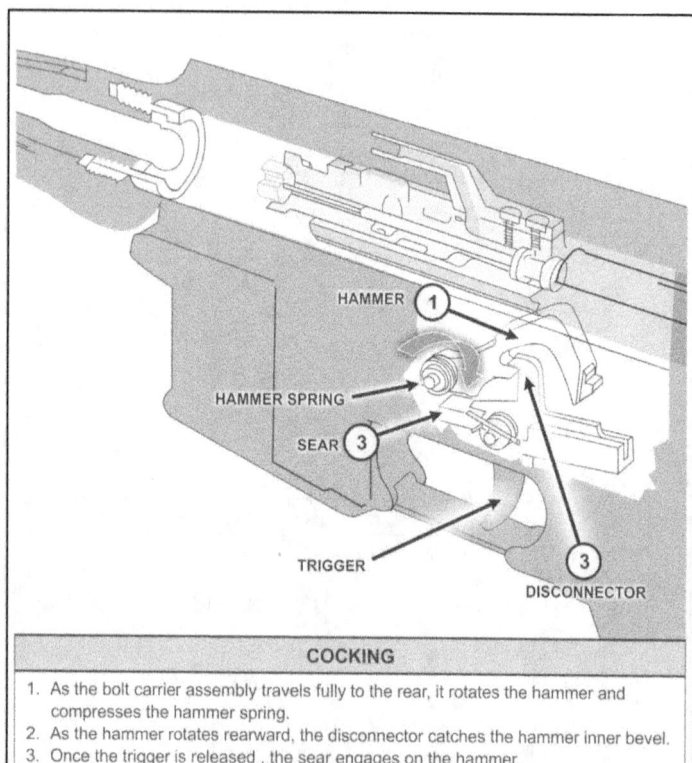

COCKING

1. As the bolt carrier assembly travels fully to the rear, it rotates the hammer and compresses the hammer spring.
2. As the hammer rotates rearward, the disconnector catches the hammer inner bevel.
3. Once the trigger is released , the sear engages on the hammer.

Figure 2-10. Cocking example

Rifle and Carbine Principles of Operation

COOLING

2-19. Cooling is the process of dissipating heat from the weapon during firing. Although not part of the cycle of functioning, cooling the weapon during firing is critical to ensure the weapon continues to operate efficiently. Firing a round generates heat and pressure within the chamber and bore, which radiates outward through the metal of the barrel.

2-20. The temperature generated by the burning of propellant powders is over one thousand degrees Fahrenheit. Some of the heat produced during firing is retained in the chamber, bore, and barrel during firing and poses a significant hazard to the firer.

2-21. How this heat is absorbed by the weapon and dissipated or removed, is a function of engineering and design. Lightweight weapons like the M4 and M16 do not have sufficient mass to withstand thermal stress efficiently. The weapon system must have a means to radiate the heat outward, away from the barrel to allow continuous firing.

2-22. There are three methods to reduce the thermal stress on a weapon. The M4- and M16-series of weapons use all three of these methods to varying degrees to cool the chamber, bore, and barrel to facilitate continuous operation. These methods of cooling are—

- **Radiational cooling** – allows for the dissipation of heat into the surrounding cooler air. This is the least efficient means of cooling, but is common to most small arms weapons, including the rifle and carbine.
- **Conduction cooling** – occurs when a heated object is in direct physical contact with a cooler object. Conduction cooling on a weapon usually results from high chamber operating temperatures being transferred into surrounding surfaces such as the barrel and receiver of the weapon. The transfer from the chamber to the cooler metals has the net effect of cooling the chamber. Thermal energy is then carried away by other means, such as radiant cooling, from these newly heated surfaces.
- **Convection cooling** – requires the presence of a moving air current. The moving air has greater potential to carry away heat. The hand guards and ARS of the rifle and carbine are designed to facilitate air movement. The heat shield reflects heat energy away from the hand guard and back towards the barrel. The net effect is an updraft that brings the cooler air in from the bottom. This process establishes a convection cycle as heated air is continually replaced by cooler air.

2-23. Soldiers should be aware of the principles of the weapon's cooling methods' direct effects on their line of sight when viewing a target through an aiming device. Dissipating heat along the length of the barrel can create a mirage effect within the line of sight which can cause a significant error to the true point of aim when using magnified optics.

Chapter 3
Aiming Devices

Every weapon has a method of aiming, that is either fixed or attached to operate the weapon effectively. Soldiers must be familiar with the various aiming devices, how they operate, and how to employ them correctly to maximize their effectiveness. This chapter provides the principles of operation of the most widely available aiming devices, and provides general information concerning their capabilities, function and use.

3-1. An aiming device is used to align the Soldier, the weapon, and the target to make an accurate and precise shot. Each aiming device functions in a different manner. To employ the weapon system to its fullest capability, the Soldier must understand how their aiming devices function.

3-2. The following aiming devices are described within this chapter:
- **Iron**. Iron represent the various types of mechanical sighting systems available on the weapon. They are available in two distinct types:
  - Iron sights (rear aperture and front sight post).
  - Back up iron sights (BUIS).
- **Optics**. These are optics predominantly for day firing, with limited night capability. The optics found within this manual come in two types:
  - Close Combat Optic (CCO).
  - Rifle Combat Optic (RCO, previously referred to as the Advanced Combat Optic Gunsight or ACOG).
- **Thermal**. These are electronic sighting systems that provide a view of the field of view (FOV) based on temperature variations. There are numerous variants of thermal optics, but are grouped into one type:
  Thermal Weapon Sight (TWS).
- **Pointer/Illuminator/Laser**. These aiming devices use either a laser beam, flood light, or other light to aim the weapon at the target. There are three types of pointers, illuminators, and lasers used by the service rifle:
  - Advanced Target Pointer Illuminator Aiming Light (ATPIAL).
  - Dual Beam Aiming Laser–Advanced (DBAL-A2).
  - Illuminator, Integrated, Small Arms (STORM).

Chapter 3

UNITS OF ANGULAR MEASUREMENT

3-3. There are two major units of angular measurement the Army uses: mils and minutes of angle (MOA). These two different units are commonly used terms to describe a measurement of accuracy when firing a weapon, system, or munition. They typically include the accuracy of a specific weapon, the performance of ammunition, and the ability of a shooter as it relates to firing the weapon.

MINUTE OF ANGLE

3-4. A minute of angle (MOA) is an angular unit of measurement equal to $1/60^{th}$ of a degree (see figure 3-1). The most common use of MOA is when describing the distance of change required when zeroing a weapon.

3-5. One MOA equals 1.047 inches per 100 yards. For most applications, a Soldier can round this to 1 inch at 100 yards or 1.1 inches at 100 meters to simplify their arithmetic.

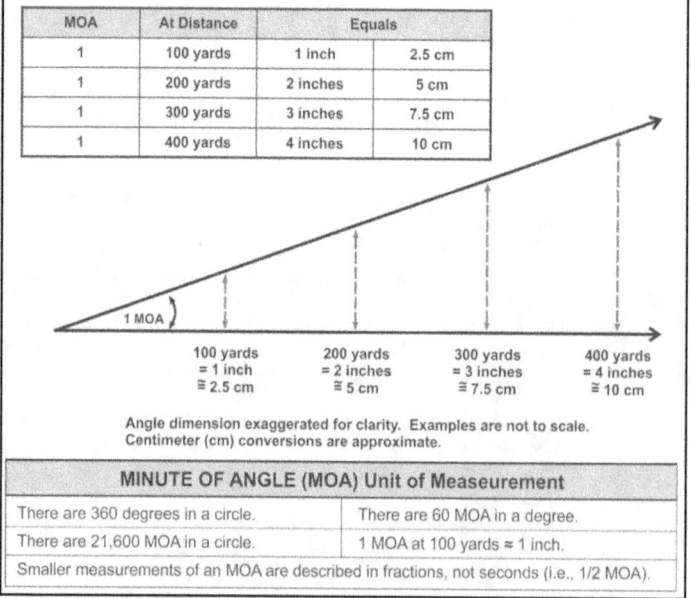

Figure 3-1. Minute of angle example

Aiming Devices

MILS

3-6. The mil is a common unit of angular measurement that is used in direct fire and indirect fire applications. (see figure 3-2)

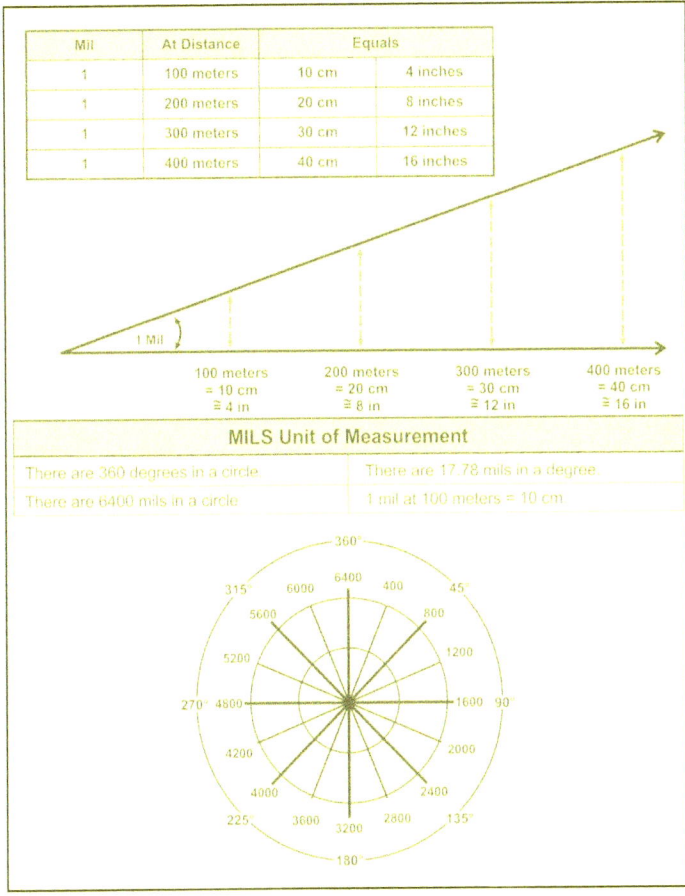

Figure 3-2. Mil example

Chapter 3

3-7. This mil to degree relationship is used when describing military reticles, ballistic relationships, aiming devices, and on a larger scale, map reading and for indirect fire.

RETICLE

3-8. A reticle is a series of fine lines in the eyepiece of an optic, such as a CCO, TWS, or RCO (see figure 3-3) used as a measuring scale with included aiming or alignment points. Reticles use either mils or minute of angle for their unit of measurement.

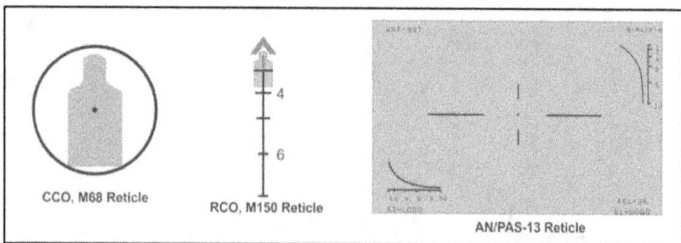

Figure 3-3. Close combat optic / Rifle combat optic reticle / Thermal reticle examples

Aiming Devices

STADIA RETICLE (STADIAMETRIC RETICLE)

3-9. Commonly used in the thermal weapon sight, a stadia reticle provides a means of rapidly determining the approximate range to target of a viewed threat, based on its standard dimensions. The stadia reticle (sometimes referred to as "stadiametric" or "choke sight") can provide approximate range to target information using width or height of a viewed dismounted target using standard threat dimensions (see figure 3-4).

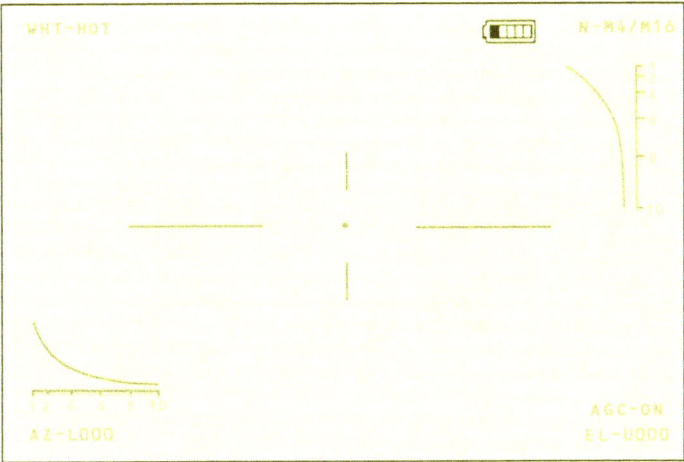

Figure 3-4. Stadia reticle example

3-10. There are two stadia reticles found on the rifle / carbine reticle within the thermal weapons sight; vertical and horizontal.
- **Vertical stadia.** At the lower left of the sight picture, Soldiers can evaluate the range to target of a standing dismounted threat.
- **Horizontal stadia.** In the upper right portion of the sight picture, Soldiers can evaluate the range to target of an exposed dismounted threat based on the width of the target.

Electromagnetic Spectrum

3-11. A major concern for the planning and use of thermal and other optics to aid in the detection process is understanding *how they function*, but more appropriately, what they can "see". Each device develops a digital representation of the scene or view it is observing based on what frequencies or wavelengths it can detect within the electromagnetic spectrum. (Note: Thermal devices see differences in heat.)

- **Thermal optics.** This equipment operates in the mid- and far-wavelength of the infrared band, which is the farthest of the infrared wavelengths from visible light. Thermal optics cannot translate ("see") visible light. Thermal optics cannot "see" infrared equipment such as infrared (IR) strobe lights, IR chemical lights, illuminators, or laser pointers. They can only identify emitted radiation in the form of heat (see figure 3-5 on page 3-7).
- **Image intensifiers (I2).** This equipment, such as night vision devices, use the near area of the infrared spectrum closest to the frequencies of visible light, as well as visible light to create a digital picture of the scene. These systems cannot "see" or detect heat or heat sources.

3-12. These sights generally operate on the principles of convection, conduction, and radiation (mentioned in chapter 2 of this publication). The sight "picks up" or translates the IR wavelength (or light) that is emitted from a target scene through one of those three methods.

3-13. Things to be aware of (planning considerations) with these optics are that they have difficulty imaging through the following:

- **Rain** – absorbs the IR emitted by the target, makes it difficult to see.
- **Water** – acts as a mirror and generally reflects IR, providing a false thermal scene.
- **Glass** – acts similar to water, interfering with the sensor's ability to accurately detect emitted radiation behind the glass.

3-14. Situations where IR can see better are the following:

- **Smoke** – will not obscure a target unless the chemical obscurant is extremely hot and dense, or if the target is sitting on top of the smoke source.
- **Dust** – may interfere with the accurate detection of the emitted thermal signature due to dust and debris density between the sensor and the target scene. Dust typically does not obscure the IR signature unless its temperature is similar to the target's.

3-15. Figure 3-5 depicts the areas of the electromagnetic spectrum. It details the various wavelengths within the spectrum where the aiming devices, night vision devices, and equipment operate. It illustrates where these items can and cannot "see" the others, respectively, within their operating range.

Aiming Devices

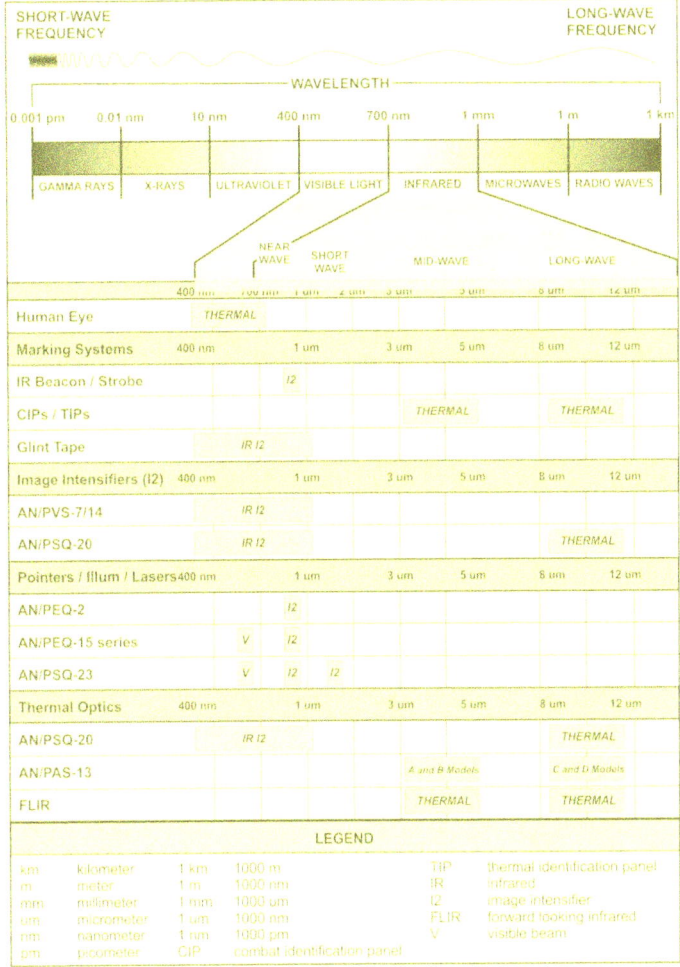

Figure 3-5. Electromagnetic spectrum

Chapter 3

OPTICS

3-16. Optics are sighting aids for rifles and carbines that provide enhanced aim point reticles, and may include magnified fields of view. Optics are specific to day operations, although may be used during limited visibility or night operations. They do not have any method of enhancing low light conditions.

3-17. Optics enhance the Soldier's ability to engage targets accurately and at extended ranges (see figure 3-6 on page 3-9). The available optics for mounting on the M4- and M16-series modular weapon system are:
- Iron Sight.
- Back Up Iron Sight (BUIS).
- CCO, M68.
- RCO, M150.

IRON SIGHT

3-18. Some versions of the M4 and M16 come with a carrying handle with an integrated rear aperture. The carrying handle may or may not be removable, depending on the version of the service rifle.

3-19. The integrated rear aperture includes adjustments for both azimuth (wind) and elevation. Specific instructions for zeroing these aiming devices are found in the respective weapon's technical manual.

3-20. The carrying handle has two selectable apertures for the engagement situation:
- Small aperture. Used for zeroing procedures and for mid- and extended-range engagements.
- Large aperture. Used during limited visibility, close quarters, and for moving targets at close or mid-range.

3-21. The iron sight uses the fixed front sight post to create the proper aim. Soldiers use the front sight post centered in the rear aperture. The following information is extracted from the weapon's technical manual.

Aiming Devices

	CARRYING HANDLE	
	DIMENSIONS	
	LENGTH	7.3 in	18.5 cm
	WIDTH	3.5 in	9.0 cm
	HEIGHT	1.9 in	4.8 cm
	WEIGHT	20.8 oz	590 g

FUNCTION	RIFLE	ADJUSTMENTS
ZERO WINDAGE	M16A2	Center rear sight aperture for mechanical zero windage
	M16A4	
	M4	
	M4A1	
ZERO ELEVATION	M16A2	300 meter mark +1 click up for 25 m zeroing. Once zeroing is complete, rotate elevation knob -1 click down to apply 300 m zero
	M16A4	
	M4	
	M4A1	
WINDAGE	M16A2	1/2 MOA
	M16A4	1/2 MOA
	M4	1 MOA
	M4A1	1 MOA
ELEVATION (RANGE) FRONT SIGHT POST	M16A2	1 1/2 MOA
	M16A4	1 1/2 MOA
	M4	1 7/8 MOA
	M4A1	1 7/8 MOA
LEGEND		
BDC bullet drop compensator	g grams	MOA minute of angle
cm centimeters	in inches	oz ounces

Figure 3-6. Carrying handle with iron sight example

Chapter 3

BACK UP IRON SIGHT

3-22. The BUIS is a semi-permanent flip-up sight equipped with a rail-grabbing base. The BUIS provides a backup capability effective out to 600 meters and can be installed on M16A4 rifles and M4-series carbines. (See figure 3-7.)

3-23. The BUIS on the first notch of the integrated rail, nearest to the charging handle. The BUIS remains on the modular weapon system (MWS) unless the carrying handle/sight is installed. The following information is extracted from the weapon's technical manual.

BACK UP IRON SIGHT (BUIS)		
DIMENSIONS		
LENGTH	2.1 in	5.3 cm
WIDTH	1.3 in	3.3 cm
HEIGHT	1.5 in	3.8 cm
WEIGHT	4.3 oz	122 g

FUNCTION		SINGLE CLICK
ZERO WINDAGE	M16A4	White Line
	M4	White Line
	M4A1	White Line
ZERO ELEVATION	M16A4	White Line
	M4	300 meter setting
	M4A1	300 meter setting
WINDAGE	M16A4	1/2 MOA
	M4	3/4 MOA
	M4A1	3/4 MOA
ELEVATION (RANGE) FRONT SIGHT POST	M16A4	1 1/2 MOA
	M4	2 MOA
	M4A1	2 MOA
LEGEND		
cm centimeters	in inches	oz ounces
g grams	MOA minute of angle	

Figure 3-7. Back up iron sight

Aiming Devices

CLOSE COMBAT OPTIC, M68

3-24. The close combat optic (CCO), M68 is a non-telescopic (unmagnified) reflex sight that is designed for the "eyes-open" method of sighting (see figure 3-8). It provides Soldiers the ability to fire with one or two eyes open, as needed for the engagement sequence in the shot process.

3-25. The CCO provides a red-dot aiming point using a 2 or 4 MOA diameter reticle, depending on the variant. The red dot aiming point follows the horizontal and vertical movement of the firer's eye, allowing the firer to remain fixed on the target. No centering or focusing on the front sight post is required. There are three versions of the CCO available in the force.

Note. Re-tighten the torque-limiting knob after firing the first three to five rounds to fully seat the M68.

3-26. The CCO is zeroed to the weapon. It must remain matched with the same weapon, attached at the same slot in the attached rail system or be re-zeroed. If the CCO must be removed for storage, Soldiers must record the serial number and the rail slot to retain zero.

Note. The weapon must be re-zeroed if the CCO is not returned to the same rail slot on the adaptive rail system.

Advantages

3-27. The CCO offers a distinct speed advantage over iron sights in most if not all engagements. The adjustments on brightness allow the Soldier to have the desired brightness from full daylight to blackout conditions.

3-28. The CCO is the preferred optic for close quarter's engagements.

Disadvantages

3-29. The CCO lacks a bullet drop compensator or other means to determine accurate range to target beyond 200m.

3-30. The following information is an extract from the equipment's technical manual for Soldier reference.

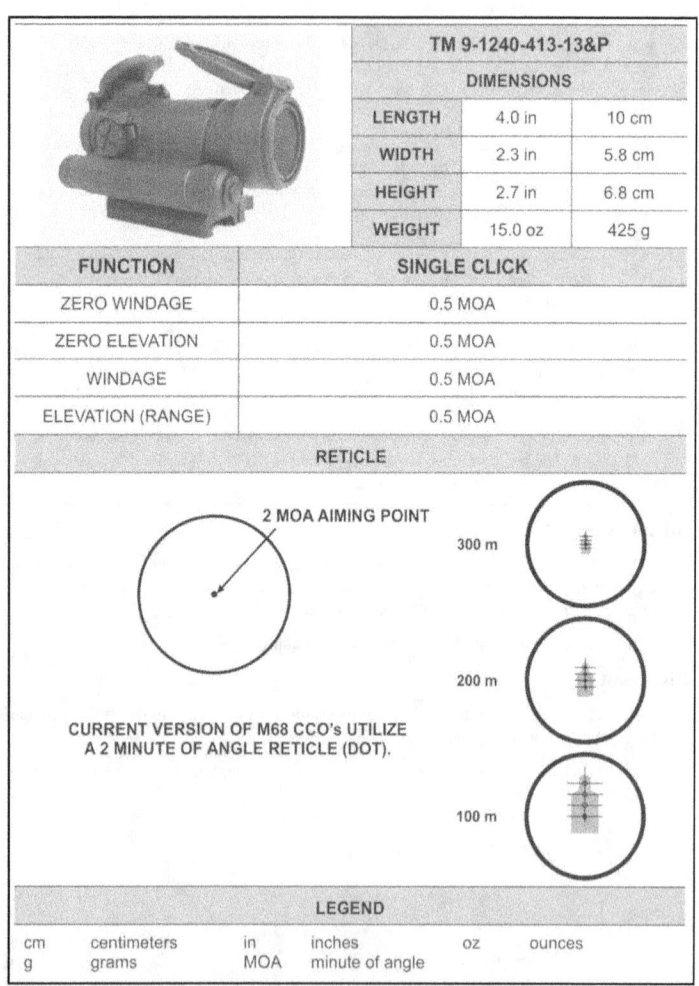

Figure 3-8. CCO Reticle, Comp M2 examples

Rifle Combat Optic

3-31. The RCO (see figure 3-9) is designed to provide enhanced target identification and hit probability for the M4-, M4A1- or M16-series weapon.

3-32. There are several versions of the RCO available for use across the force. Soldiers must be familiar with their specific version of their assigned RCO, and be knowledgeable on the specific procedures for alignment and operation (see figure 3-9 for RCO azimuth and elevation adjustments).

3-33. The reticle pattern provides quick target acquisition at close combat ranges to 800 meters using the bullet drop compensator (BDC) (see figure 3-10 on page 3-15). It is designed with dual illuminated technology, using fiber optics for daytime employment and tritium for nighttime and low-light use.

3-34. The RCO is a lightweight, rugged, fast, and accurate 4x power optic scope specifically designed to allow the Soldier to keep both eyes open while engaging targets and maintain maximum situational awareness.

Advantages

3-35. The bullet drop compensator (BDC) is accurate for extended range engagements using either M855 or M855A1 ball ammunition. The ballistic difference between the two rounds is negligible under 400 meters and requires no hold determinations.

3-36. This is a widely fielded optic that is rugged, durable, and operates in limited light conditions. The self-illuminating reticle allows for continuous operations through end evening nautical twilight (EENT).

Disadvantages

3-37. This optic's ocular view is limited when engaging targets in close quarters engagements. This requires additional training to master the close quarter's skills while employing the RCO to achieve overmatch against the threat.

3-38. The RCO reticle does not include stadia lines. Windage must be applied by the shooter from a determined estimate. The RCO has a specific eye relief of 70-mm (millimeter) or 1.5 inches. If the eye relief is not correct, the image size will be reduced.

3-39. The fiber optic illuminator element can provide excessive light to the reticle during certain conditions that produce a glare. The RCO does not have a mechanical or built in method to reduce the effects of the glare created. The increased lighting may interfere with the shooter's point of aim and hold determinations. Soldiers may use alternate methods to reduce the glare by reducing the amount of fiber optic exposed to direct sunlight during operating conditions.

3-40. The following information is an extract from the equipment's technical manual for Soldier reference.

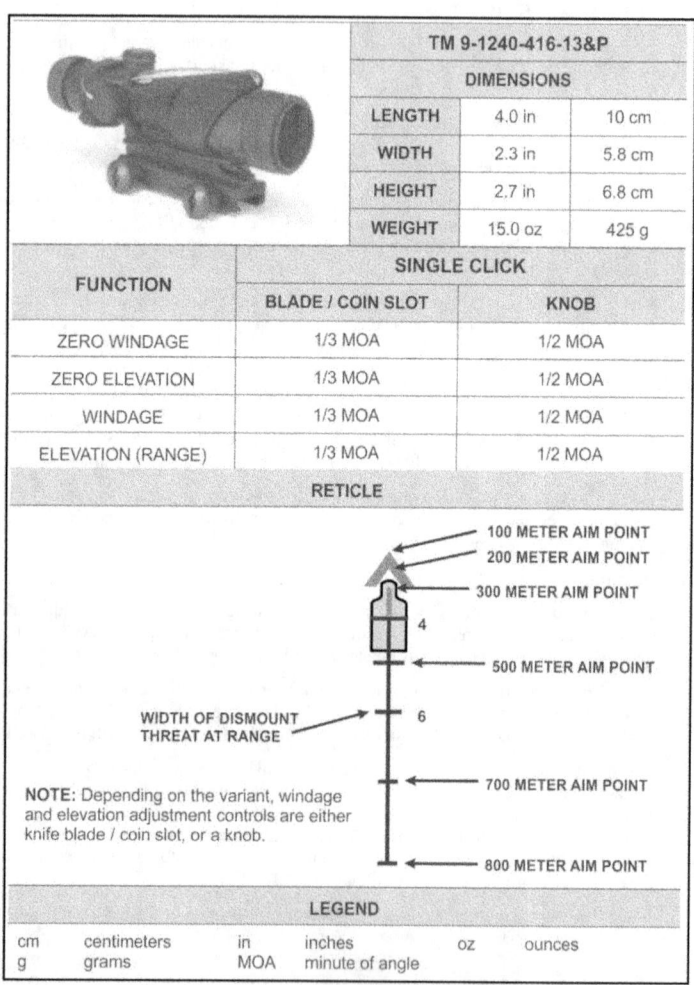

Figure 3-9. RCO reticle example

Aiming Devices

THERMAL SIGHTS

3-41. Thermal sights are target acquisition and aiming sensors that digitally replicate the field of view based on an estimation of the temperature. They use advanced forward-looking infrared technology that identify the infrared emitted radiation (heat) of a field of view, and translate those temperatures into a gray- or color-scaled image. The TWS is capable of target acquisition under conditions of limited visibility, such as darkness, smoke, fog, dust, and haze, and operates effectively during the day and night.

3-42. The TWS is composed of five functional groups: (See figure 3-10.)

- **Objective lens** – receives IR light emitting from an object and its surroundings. The objective lens magnifies and projects the IR light.
- **Detector assembly** – senses the IR light and coverts it to a video signal.
- **Sensor assembly** – the sensor electronics processes the video for display on the liquid crystal display (LCD) array in the field of view.
- **LCD array/eyepiece** – the LCD array provides the IR image along with the reticle selected. The light from the LCD array is at the eyepiece.
- **User controls** – the control electronics allows the user to interface with the device to adjust contrast, thermal gain, sensitivity, reticle display, and magnification.

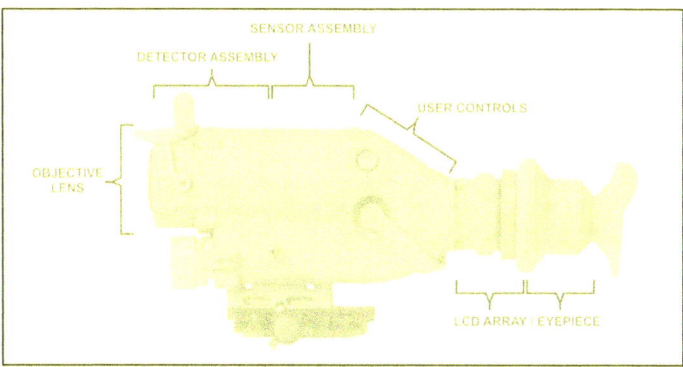

Figure 3-10. Thermal weapon sight example

3-43. A small detector used in thermal sensors or optics to identify IR radiation with wavelengths between 3 and 30 µm (micrometer). The thermal optic calculates and processes the thermal scene into a correlating video image signal based on the temperature identified. These optics can differentiate thermal variations of 1 degree Celsius of the viewable scene. These variations generate a corresponding contrasting gradient that develops a thermal representation on the LCD screen in the eyepiece.

AN/PAS-13 Series of Weapon Thermal Sights

3-44. There are several versions of weapons thermal sights (WTS) available for use across the force. Soldiers must be familiar with their specific model and version of their assigned weapon thermal sight, and be knowledgeable on the specific procedures for alignment and operation. The various models and versions are identified in their official model nomenclature:

- **Version 1 (v1)** – Light Weapons Thermal Sight (LWTS).
- **Version 2 (v2)** – Medium Weapons Thermal Sight (MWTS).
- **Version 3 (v3)** – Heavy Weapons Thermal Sight (HWTS).

3-45. Weapons thermal sights are silent, lightweight, and compact, and have durable battery-powered IR imaging sensors that operate with low battery consumption. (See figure 3-11.)

Advantages

3-46. Military grade weapon thermal weapon sights are designed with the following advantages:

- Small and lightweight.
- Real-time imagery. Devices provide real-time video of the thermal scene immediately after power on.
- Long-lasting battery life. Low power consumption over time.
- Reliable. Long mean time between failures (MTBF).
- Quiet. The lack of a cooling element allows for a very low operating noise level.
- One optic fits on multiple weapons. The use of the ARS rail mounting bracket allows for the same optic to be used on other weapons.
- The F- and G-models attach in front of other aiming devices to improve their capabilities and eliminate the zeroing procedures for the device.

Disadvantages

3-47. These devices have limitations that Soldiers should take into consideration, particularly during combat operations. The primary disadvantages are:

- Cannot interpret ("see") multispectral infrared. These systems view a specific wavelength for emitted radiation (heat variations), and do not allow viewing of all aiming and marking devices at night.
- Reliance on rechargeable batteries and charging stations. Although the batteries are common and have a relatively long battery life, additional equipment is required to charge them. If common nonrechargeable (alkaline) batteries are used, a separate battery adapter is typically required.
- Cannot interpret thermal signatures behind glass or water effectively.
- Thermal systems cannot always detect friendly marking systems worn by dismounts.

Aiming Devices

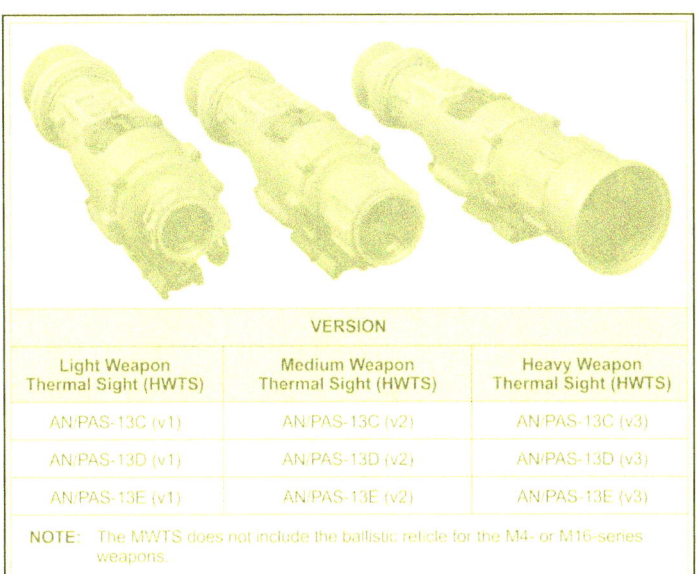

VERSION		
Light Weapon Thermal Sight (HWTS)	Medium Weapon Thermal Sight (HWTS)	Heavy Weapon Thermal Sight (HWTS)
AN/PAS-13C (v1)	AN/PAS-13C (v2)	AN/PAS-13C (v3)
AN/PAS-13D (v1)	AN/PAS-13D (v2)	AN/PAS-13D (v3)
AN/PAS-13E (v1)	AN/PAS-13E (v2)	AN/PAS-13E (v3)
NOTE: The MWTS does not include the ballistic reticle for the M4- or M16-series weapons.		

Figure 3-11. Weapon thermal sights by version

3-48. Thermal sight has a wide field of view and a narrow field of view (see figures 3-12 and 3-13).

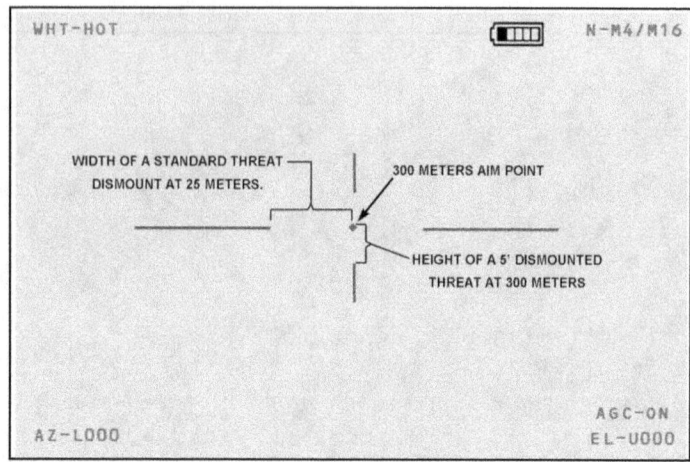

Figure 3-12. Thermal weapons sight, narrow field of view reticle example

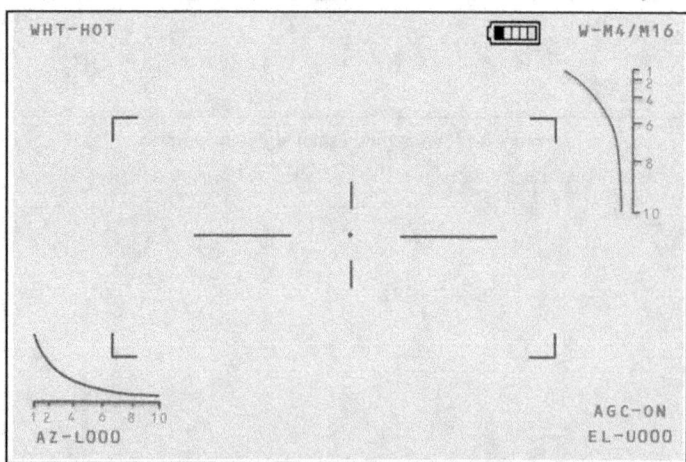

Figure 3-13. Thermal weapons sight, wide field of view reticle example

Aiming Devices

POINTERS / ILLUMINATORS / LASERS

3-49. Pointers, illuminators, and laser devices for small arms weapons emit a collimated beam of IR light for precise aiming and a separate IR beam for illumination. These devices operate in one single mode at a time, as selected by the user. The laser is activated by a selector switch on the device or by a remote mechanism installed on the weapon. The basic two modes or functions are:

- **Pointer.** When used as a pointer or aiming device, a small, pin-point beam is emitted from the device. The IR beam provides an infrared visible point when it strikes an object or target. The IR beam operates in the 400 to 800 nanometer wavelength and can only be seen by 12 optics, such as the AN-PVS-7 or -14 night vision devices.
- **Illuminator.** Typically used to illuminate a close quarters area as an infrared flood light. The illuminator provides a flood-light effect for the Soldier when used in conjunction with 12 night vision devices.

Note. Laser is an acronym for light amplified stimulated emitted radiation, but is predominantly used as a proper noun.

3-50. The following devices (see table 3-1) are the most common laser pointing devices available for use on the M4- and M16-weapons.

Table 3-1. Laser Aiming Devices for the M4 and M16

Laser Aiming Device	Device Name	Reference
AN/PEQ-2	Target Pointer/Illuminator/Aiming Light (TPIAL)	TM 9-5855-1915-13&P
AN/PEQ-15	Advanced Target Pointer/Illuminator/Aiming Light (ATPIAL)	TM 9-5855-1914-13&P
AN/PEQ-15A	Dual Beam Aiming Laser – Advanced2 (DBAL-A2)	TM 9-5855-1912-13&P
AN/PSQ-23	Illuminator, Integrated, Small Arms (STORM)	TM 9-5855-1913-13&P

Note. The ATPIAL, DBAL-A2, and STORM have collocated IR and visible aiming lasers. A single set of adjusters move both aiming beams. Although the aiming lasers are collocated, Soldiers should zero the laser they intend to use as their primary pointer to ensure accuracy and consistency during operation.

AN/PEQ-2 Target Pointer/Illuminator Aiming Light (TPIAL)

3-51. AN/PEQ-2 aiming devices are Class IIIb laser devices that emit a collimated beam of IR light for precise aiming and a separate IR beam for illumination of the target or target area (see figure 3-14 on page 3-21). Both beams can be independently zeroed to the weapon and to each other. The beams can be operated individually or in combination in both high and low power settings.

> *Note.* The IR illuminator is equipped with an adjustable bezel to vary the size of the illumination beam based on the size and distance of the target.

3-52. The aiming devices are used with night observation devices (NODs) and can be used as handheld illuminators/pointers or mounted on the weapon with the included brackets and accessory mounts. In the weapon-mounted mode, the aiming devices can be used to direct fire and to illuminate and designate targets.

3-53. The aiming light is activated by pressing on either the ON/OFF switch lever, or the button on the optional cable switch. Either switch connects power from two AA batteries to an internal electronic circuit which produces the infrared laser. Internal lenses focus the infrared light into a narrow beam. The direction of the beam is controlled by rotating the mechanical Adjusters with click detents. These adjusters are used to zero the aiming light to the weapon.

3-54. Once zeroed to the weapon, the aiming light projects the beam along the line of fire of the weapon. The optical baffle prevents off-axis viewing of the aiming light beam by the enemy.

> **CAUTION**
>
> A safety block is provided for training purposes to limit the operator from selecting high power modes of operation.

3-55. The following information is an extract from the equipment's technical manual for Soldier reference.

Aiming Devices

	TM 9-5855-1915-13&P		
	DIMENSIONS		
	LENGTH	6.4 in	16.3 cm
	WIDTH	2.8 in	7.1 cm
	HEIGHT	1.2 in	3 cm
	WEIGHT	9.5 oz	269 g
POWER				
BATTERY LIFE	100 hours >32°		
	36 hours <32°		
POWER SOURCE	2 each AA batteries		
MODE OF OPERATION				
MODE	MARKINGS	TGT LASER	ILLUM LASER
0	OFF	OFF	OFF
1	AIM LO	LOW POWER	OFF
2	DUAL LO	LOW POWER	LOW POWER
3	AIM HI	HIGH POWER	OFF
4	DUAL LO/HI	HIGH POWER	LOW POWER
5	DUAL HI	HIGH POWER	HIGH POWER
LASER	DIVERGENCE	WAVELENGTH	
IR BEAM	0.3 mRad	820-850 nm	
IR ILLUMINATOR	3.0 mRad	820-850 nm	
LEGEND				
cm centimeters	IR infrared	oz ounces	
g grams	mRad milliradians		
in inches	nm nanometers		

Figure 3-14. AN/PEQ-2

AN/PEQ-15 Advanced Target Pointer/Illuminator/Aiming Light

3-56. The AN/PEQ-15 ATPIAL is a multifunctional laser that emits both a visible and IR light for precise weapon aiming and target/area illumination. This ruggedized system can be used as a handheld illuminator/pointer or can be mounted to weapons equipped with an M4- or M5-ARS (Military Standard [MIL STD] 1913).

- **Visible light** – can be used to boresight the device to a weapon without the need of night vision goggles. A visible red-dot aiming laser can also be selected to provide precise aiming of a weapon during daylight or night operations.
- **Infrared laser** – emit a highly collimated beam of IR light for precise weapon aiming. A separate IR-illuminating laser can be adjusted from a flood light mode to a single point spot-divergence mode.

3-57. The lasers can be used as handheld illuminator pointers, or can be weapon-mounted with included hardware. The co-aligned visible and IR aiming lasers emit through laser ports in the front of the housing. These highly capable aiming lasers allow for accurate nighttime aiming and system boresighting.

3-58. The AN/PEQ-15 has an integrated rail grabber molded into the body to reduce weight and additional mounting hardware. (Refer to TM 9-5855-1914-13&P for more information.)

> **CAUTION**
>
> The AN/PEQ-15 can be used during force-on-force training in the low power modes only. High power modes can be used on live-fire ranges exceeding 220 meters only.

3-59. The AN/PEQ-15, ATPIAL's (see figure 3-15 on page 3-23) visible aiming laser provides for active target acquisition in low light conditions and close-quarters combat situations, and allows users to zero using the borelight without using NOD. When used in conjunction with NODs, its IR aiming and illumination lasers provide for active, covert target acquisition in low light or complete darkness.

3-60. The ATPIAL visible and IR aiming lasers are co-aligned. A single set of adjusters moves both aiming beams, and the user can boresight/zero using either aiming laser. The following information is an extract from the equipment's technical manual for Soldier reference.

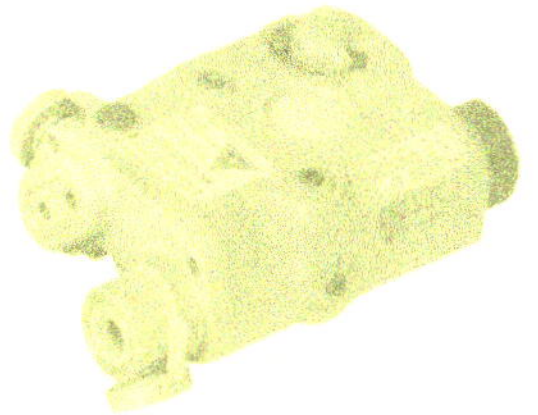

Aiming Devices

TM 9-5855-1914-13&P		
DIMENSIONS		
LENGTH	4.6 in	11.7 cm
WIDTH	2.8 in	7.1 cm
HEIGHT	1.9 in	4.1 cm
WEIGHT	7.5 oz	213 g
POWER		
BATTERY LIFE	>6 hours in DUAL HIGH (DH) mode	
POWER SOURCE	1 each DL-123A, 3 volt	

MODE OF OPERATION		
POSITION	MODE	REMARKS
VIS AL	Vis Aiming Laser	Visible Aim Laser ON
O	OFF	Prevents inadvertent laser burst
P	Program	Sets the desired IR pulse rate
AL	AIM LOW	Low power of Aiming Laser
DL	DUAL LOW	Aiming Laser and Illuminator on LOW
AH	AIM HIGH	Aiming Laser set to HIGH
IH	ILLUM HIGH	IR Illuminator set to HIGH
DH	DUAL HIGH	IR Aim and Illuminator set to HIGH

LASER	DIVERGENCE	WAVELENGTH
IR BEAM	0.5 mRad	820-850 nm
IR ILLUMINATOR	1.0 to 105 mRad	820-850 nm
VISIBLE AIMING	0.5 mRad	605-665 nm

LEGEND					
cm	centimeters	IR	infrared	oz	ounces
g	grams	mRad	milliradians		
in	inches	nm	nanometers		

Figure 3-15. AN/PEQ-15, ATPIAL

AN/PEQ-15A, Dual Beam Aiming Laser – Advanced2

3-61. The AN/PEQ-15A DBAL-A2 is a multifunctional laser device that emits IR pointing and illumination light, as well as a visible laser for precise weapon aiming and target/area illumination. The visible and IR aiming lasers are co-aligned enabling the visible laser to be used to boresight both aiming lasers to a weapon without the need for night vision devices. This ruggedized system can be used as a handheld illuminator/pointer or can be mounted to weapons equipped with an M4 or M5 adapter rail system (MIL-STD-1913).

- **Visible light** – can be used to boresight the device to a weapon without the need of night vision goggles. A visible red-dot aiming laser can also be selected to provide precise aiming of a weapon during daylight or night operations.
- **Infrared laser** – emits a tightly focused beam of IR light for precise aiming of the weapon. A separate IR illumination provides supplemental IR illumination of the target or target area. The IR illuminator is equipped with an adjustable bezel to vary the size of the illumination beam on the size and distance to the target (flood to point divergence).

3-62. The lasers can be used as hand-held illuminator pointers, or can be weapon-mounted with included hardware. These highly capable aiming lasers allow for accurate nighttime aiming and system boresighting.

3-63. The AN/PEQ-15A, DBAL-A2 (see figure 3-16 on page 3-25) visible aiming laser provides for active target acquisition in low light conditions and close quarters combat situations, and allows users to zero using the borelight without using NODs. When used in conjunction with NODs, its IR aiming and illumination lasers provide for active, covert target acquisition in low light or complete darkness.

3-64. The DBAL-A2 visible and IR aiming lasers are co-aligned. A single set of adjusters moves both aiming beams, and the user can boresight/zero using either aiming laser. The following information is an extract from the equipment's technical manual for Soldier reference.

Aiming Devices

TM 9-5855-1912-13&P		
DIMENSIONS		
LENGTH	3.5 in	8.7 cm
WIDTH	2.9 in	7.4 cm
HEIGHT	1.9 in	4.8 cm
WEIGHT	8 oz	224 g

POWER	
BATTERY LIFE	>5.5 hours in IR DUAL HIGH mode
POWER SOURCE	1 each DL-123A, 3 volt

MODE OF OPERATION		
POSITION	MODE	REMARKS
AL	LOW POWER	Low power for aim laser
AH	HIGH POWER	High power for aim laser
VIS A	VIS AIM RED	Aiming or marking laser for daylight
VIS A	VIS AIM GREEN	Aiming or marking laser for daylight

LASER	DIVERGENCE	WAVELENGTH
IR BEAM	0.3 mRad	840 nm
IR ILLUMINATOR	0.5 to 75 mRad	840 nm
VISIBLE AIM, RED	0.3 mRad	635 nm
VISIBLE AIM, GREEN	0.5 mRad	532 nm

LEGEND					
cm	centimeters	IR	infrared	oz	ounces
g	grams	mRad	milliradians		
in	inches	nm	nanometers		

Figure 3-16. AN/PEQ-15A, DBAL-A2

AN/PSQ-23, Illuminator, Integrated, Small Arms

3-65. The AN/PSQ-23 is a battery operated laser range finder (LRF) and digital magnetic compass (DMC) with integrated multifunctional lasers. The illuminator, integrated, small arms device is commonly referred to as the STORM laser. The visible and IR aiming lasers are co-aligned enabling the visible laser to be used to boresight both aiming lasers to a weapon without the need for night vision devices. This ruggedized system can be used as a handheld illuminator/pointer or can be mounted to weapons equipped with an M4 or M5 adapter rail system (MIL-STD-1913).

- **Laser range finder** – provides range to target information from 20 meters to 10,000 meters with an accuracy of +/- 1.5 meters.
- **Digital magnetic compass** – provides azimuth information and limited elevation information to the operator. The azimuth accuracy is +/- 0.5 degrees to +/- 1.5 degrees. The elevation accuracy is +/- 0.2 degrees. The DMC can identify bank or slopes up to 45 degrees with an accuracy of +/- 0.2 degrees.
- **Visible light** – provides for active target acquisition in low light and close quarters combat situations without the need for night vision devices. It can be used to boresight the device to a weapon without the need of night vision devices. A visible red-dot aiming laser can also be selected to provide precise aiming of a weapon during daylight or night operations.
- **Infrared laser** – emits a tightly focused beam of IR light for precise aiming of the weapon. A separate IR illumination provides supplemental IR illumination of the target or target area. The IR illuminator is equipped with an adjustable bezel to vary the size of the illumination beam on the size and distance to the target (flood to point divergence).
- **Infrared illuminator** – the STORM features a separately adjustable IR illuminator with adjustable divergence. It is fixed in the device housing and is set parallel to the rail mount.

Note. The STORM's LRF and DMC may be used in combination to obtain accurate positioning information for targeting purposes and other tactical applications.

3-66. The integrated visible aim laser (VAL) and illumination lasers provide for active, covert target acquisition in low light or complete darkness when used in conjunction with night vision devices. The STORM is also equipped with a tactical engagement simulation (TES) laser allowing it to be used in a laser-based training environment.

3-67. The AN/PEQ-15A, DBAL-A2 visible aiming laser provides for active target acquisition in low light conditions and close-quarters combat situations, and allows users to zero using the borelight without using NODs. When used in conjunction with NODs, its IR aiming and illumination lasers provide for active, covert target acquisition in low light or complete darkness. The following information is an extract from the equipment's technical manual for Soldier reference (see figure 3-17 on page 3-27).

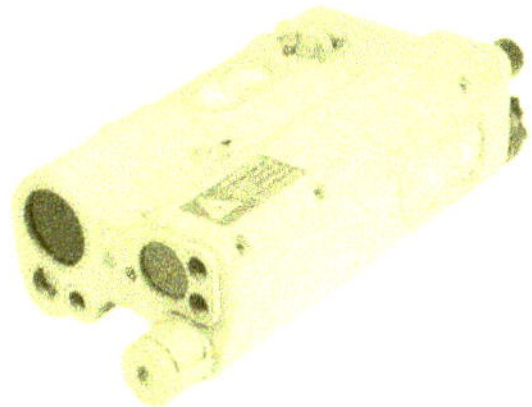

Aiming Devices

TM 9-5855-1913-13&P		
DIMENSIONS		
LENGTH	7.3 in	18.5 cm
WIDTH	3.5 in	9.0 cm
HEIGHT	1.9 in	4.8 cm
WEIGHT	20.8 oz	590 g

POWER	
BATTERY LIFE	>5.5 hours in IR DUAL HIGH mode
POWER SOURCE	2 each DL-123A, 3 volt

MODE OF OPERATION		
POSITION	MODE	REMARKS
VH	VIS HIGH	Aiming or marking in daylight/indoor
AH	AIM HIGH	IR operates on high power
IH	ILLUM HIGH	IR illum operates on high power
DH	DUAL HIGH	IR/Illum both operate on high power
BUTTON	MODE	REMARKS
L	Laser activate	Activates aiming laser
R	Range/Compass	Press/Hold 3 sec to enter menu power

LASER	DIVERGENCE	WAVELENGTH
IR BEAM	0.5 mRad	820-850 nm
IR ILLUMINATOR	1.0 to 100 mRad	820-850 nm
VISIBLE AIM, RED	0.5 mRad	605-665 nm
LASER RANGE FINDER	1.0 mRad	1570 nm

LEGEND					
cm	centimeters	IR	infrared	oz	ounces
g	grams	mRad	milliradians		
in	inches	nm	nanometers		

Figure 3-17. AN/PSQ-23, STORM

This page intentionally left blank.

Chapter 4

Mountable Equipment

Both the M4- and M16-series of weapons have a wide variety of attachments to increase Soldier lethality, situational awareness, and overmatch. The attachments can be applied in various locations on the weapon system. Soldiers must understand what the attachments are, how they are correctly positioned, how to align them with the weapon system, and how to integrate them into use to maximize the system's capabilities.

This chapter explains how the ARS is used to mount the various attachments. It describes the weapons, aiming devices, and accessories available for mounting, and includes general information on the proper mounting location as well as their basic capabilities.

ADAPTIVE RAIL SYSTEM

4-1. The ARS and rail grabbers are designed for M16- and M4-/M4A1-series weapons to mount:
- Weapons.
- Aiming devices.
- Accessories.

4-2. The ARS provides a secure mounting point for various accessories that may be mounted on the weapon's top, bottom, left, and right. Each rail groove has an incremental number identifying the slot location, starting from the rear of the weapon.

4-3. Soldiers should record the attachment or equipment's serial number (if applicable), the location of the attachment (for example, markings between lugs), and any boresight or alignment settings specific to the equipment at that location.

4-4. Once complete, the Soldier should mark the mounting bracket to identify the tightened position with a permanent marker. Marking the mounting bracket allows for rapid identification of loosening hardware during firing. Soldiers must periodically verify the mounting hardware does not loosen during operation. During zeroing or zero confirmation operations, Soldiers should retighten the mounting hardware after the first five rounds.

4-5. Soldiers must ensure the equipment is firmly affixed to the ARS before tie down is complete. If the attachments are loose, their accuracy and effectiveness will be degraded.

Chapter 4

MOUNTABLE WEAPONS

4-6. There are two types of weapons that can be physically attached to the M16-/M4-series rifles; grenade launchers and shotguns. These weapons are standard components of the unit's organizational equipment and serve specific purposes during combat operations.

4-7. These weapons are mounted under the barrel of the service rifle at specific locations. They may be removed by a qualified armorer only.

GRENADE LAUNCHERS

4-8. The M320/M320A1 grenade launcher is a lightweight grenade launcher that can operate in a stand-alone or attached configuration. The M320/M320A1 grenade launcher uses an integrated double-action-only trigger system. The M320 series is the replacement weapon for the M203. (See figure 4-1.)

Figure 4-1. M320 attached to M4 series carbine example

4-9. The M203 is a breach loaded attachable grenade launcher that is affixed to the bottom of the barrel of the M16-/M4-series rifle. The M203 cannot be used in a stand-alone configuration. (See figure 4-2)

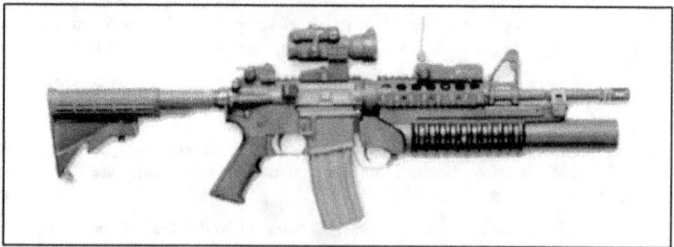

Figure 4-2. M203 grenade launcher example

Mountable Equipment

4-10. Each mountable 40mm grenade launcher provides the following capabilities to the small unit (see the appropriate TM for authorized use):
- Pyrotechnic signal and spotting rounds:
  - Star cluster, white.
  - Star parachute, white.
  - Star parachute, green.
  - Star parachute, red.
  - Smoke, yellow.
  - Smoke, green.
  - Smoke, red.
  - Illumination, infrared.
- High explosive (HE).
- High explosive, dual purpose (HEDP).
- Nonlethal.
- Training practice (TP).

SHOTGUN SYSTEM

4-11. The M26 Modular Accessory Shotgun System (MASS) is an under-barrel shotgun attachment for the M16/M4/M4A1. The M26 uses a 3- or 5-round detachable box magazine and provides Soldiers with additional tactical capabilities. (Refer to TC 3-22.12 for more information). (See figure 4-3.)

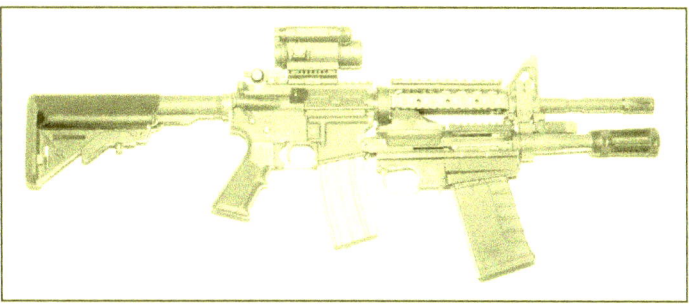

Figure 4-3. M26 shotgun example

4-12. The M26 provides specific tactical capabilities to the Soldier using the following ammunition:
- Slug, Door breaching.
- Shot range, 00 buckshot.
- Nonlethal, rubber slug, buckshot, and riot control.

Chapter 4

MOUNTABLE AIMING DEVICES

4-13. Aiming devices mounted to the weapon system should be placed in a specific location on the weapon to maximize their capabilities. Table 4-1 provides the preferred mounting locations of the most common attachments.

Table 4-1. Attachment Related Technical Manuals and Mounting

Attachment	Technical Manual	M4/M4A1, M16A4	M4/M4A1	M16A2/A3
BUIS		UR	UR	
CCO, M68	TM 9-1240-413-13&P	UR*	UR*	MT
RCO, M150	TM 9-1240-416-13&P	UR	UR	MT
AN/PVS-14	TM 11-5855-306-10	UR***		
AN/PEQ-15A	TM 9-5855-1912-13&P	RG**	BA	BA
AN/PEQ-15	TM 9-5855-1914-13&P	RG**	BA	BA
AN/PAS-13B(V1), LWTS	TM 11-5855-312-10	UR	UR	MT
AN/PAS-13B(V3), HWTS	TM 11-5855-312-10	UR	UR	MT
AN/PAS-13C(V1), LWTS	TM 11-5855-316-10	UR	UR	MT
AN/PAS-13C(V3), HWTS	TM 11-5855-316-10	UR	UR	MT
AN/PAS-13D(V)1 LWTS	TM 11-5855-324-10	UR	UR	MT
AN/PAS-13D(V2), MWTS	TM 11-5855-317-10	UR	UR	MT
AN/PAS-13D(V3), HWTS	TM 11-5855-317-10	UR	UR	MT
AN/PSQ-23	TM 9-5855-1913-13&P	RG**	BA	BA
Legend: BA – Bracket Assembly BUIS – Back up Irion Sight CCO – Close Combat Optic HTWS – Heavy Thermal Weapons Sight LTWS – Light Thermal Sight MWTS – Medium Thermal Sight MT – M16 Mount RCO – Rifle Combat Optic RG – Rail Grabber UR – Upper Receiver * With a half-moon spacer installed. ** Picatinny or Insight rail grabbers may be used. *** If used in conjunction with the CCO, the CCO will mount on the top rail of the ARS.				

Mountable Equipment

MOUNTABLE ACCESSORIES

4-14. Mountable accessories are items that may be attached to a weapon but are not required for operation. They provide assistance stabilizing the weapon or provide white-light illumination for specific tactical operations.

4-15. These devices are authorized as needed by the small unit. Some mountable accessories are aftermarket (commercial-off-the-shelf, or COTS) items that use the ARS for semipermanent attachment.

BIPOD

4-16. Bipods are highly adjustable that enhance stability within the battle space environment. They are secured by the front sling swivel or the advanced rail system on the foregrip of the weapon. They can be used in combination with a sand sock or other buttstock support to provide an extremely stable firing platform. (See figure 4-4.)

4-17. The bipod is an additional means to stabilize the weapon in various shooting positions. Despite primarily being used in prone position, bipods can be used for additional support in alternate shooting positions while using barricade supports. The bipod provides additional support which facilitates acquisition of muscle relaxation and natural point of aim. The use of bipods in barricade shooting can increase the Soldier's efficiency and probability of a first round hit while engaging targets.

Figure 4-4. Bipod example

Chapter 4

VERTICAL FOREGRIP

4-18. Vertical foregrips (VFGs) assist in transitioning from target to target in close quarter combat. (See figure 4-5.)

4-19. The further out the Soldier mounts the VFG, the smoother and quicker his transitions between multiple targets will be, however he should not mount it so far forward that using the VFG is uncomfortable.

Figure 4-5. Vertical foregrip example

FOREGRIP WITH INTEGRATED BIPODS

4-20. VFGs with integrated bipods are acceptable for common use. They combine the VFG capability with a small, limited adjustment bipod. They typically lack the full adjustment capabilities of full bipods, but provide a compact stable extrusion for the firer.

MOUNTED LIGHTS

4-21. The weapon-mounted lights are commonly issued throughout the Army. The purpose of the weapon mounted lights is to provide illumination and assist in target acquisition and identification during limited visibility operations.

4-22. Most weapon mounted lights provide selection between white light and infrared capabilities. Employment of the weapon mounted light is based upon mission, enemy, terrain and weather, troops and support available, time available, civil considerations (METT-TC) and unit SOP. The weapon mounted lights should be mounted in such a manner that the Soldier can activate and deactivate them efficiently and their placement does not hinder the use of any other attachment or accessory. They must be attached in such a manner as to prevent negligent or unintentional discharge of white light illumination during movement or climbing.

Chapter 5

EMPLOYMENT

The rifleman's primary role is to engage the enemy with well-aimed shots. (Refer to ATP 3-21.8 for more information.) In this capacity, the rate of fire for the M4 rifle is not based on how fast the Soldier can pull the trigger. Rather, it is based on how fast the Soldier can consistently acquire and engage the enemy with accuracy and precision.

Consistently hitting a target with precision is a complex interaction of factors immediately before, during, and after the round fires. These interactions include maintaining postural steadiness, establishing and maintaining the proper aim on the target, stabilization of the weapon while pressing the trigger, and adjusting for environmental and battlefield conditions.

5-1. Every Soldier must adapt to the firing situation, integrate the rules of firearms safety, manipulate the fire control, and instinctively know when, how, and where to shoot. It is directly influenced by the Soldier's ability to hit the target under conditions of extreme stress:

- Accurately interpret and act upon perceptual cues related to the target, front and rear sights, rifle movement, and body movement.
- Execute minute movements of the hands, elbows, legs, feet, and cheek.
- Coordinate gross-motor control of their body positioning with fine-motor control of the trigger finger.

5-2. Regardless of the weapon system, the goal of shooting remains constant: well-aimed shots. To achieve this end state there are two truths, Soldier's must—

- Properly point the weapon (sight alignment *and* sight picture).
- Fire the weapon without disturbing the aim.

5-3. To accomplish this, Soldiers must master sight alignment, sight picture, and trigger control.

- **Sight alignment** – sight alignment is the relationship between the aiming device and the firer's eye. To achieve proper and effective aim, the focus of the firer's eye needs to be on the front sight post or reticle. The Soldier must maintain sight alignment throughout the aiming process.
- **Sight picture** – the sight picture is the placement of the aligned sights on the target.
- **Trigger control** – the skillful manipulation of the trigger that causes the rifle to fire without disturbing the aim.

Chapter 5

SHOT PROCESS

5-4. The **shot process** is the basic outline of an individual engagement sequence all firers consider during an engagement, regardless of the weapon employed. The shot process formulates all decisions, calculations, and actions that lead to taking the shot. The shot process may be interrupted at any point before the sear disengaging and firing the weapon should the situation change.

5-5. The shot process has three distinct phases:
- **Pre-shot**.
- **Shot**.
- **Post-shot**.

5-6. To achieve consistent, accurate, well-aimed shots, Soldiers must understand and correctly apply the shot process. The sequence of the shot process does not change, however, the application of each element vary based on the conditions of the engagement.

5-7. Every shot that the Soldier takes has a complete shot process. Grouping, for example, is simply moving through the shot process several times in rapid succession.

5-8. The shot process allows the Soldier to focus on one cognitive task at a time. The Soldier must maintain the ability to mentally organize the shot process's tasks and actions into a disciplined mental checklist, and focus their attention on activities which produce the desired outcome; a well-aimed shot.

5-9. The level of attention allocated to each element during the shot process is proportional to the conditions of each individual shot. Table 5-1 provides an example of a shot process.

Table 5-1. Shot Process example

Phase	Element
Pre-shot	Position
	Natural Point of Aim
	Sight Alignment / Picture
	Hold
Shot	Refine Aim
	Breathing Control
	Trigger Control
Post-shot	Follow-through
	Recoil management
	Call the Shot
	Evaluate

Employment

FUNCTIONAL ELEMENTS OF THE SHOT PROCESS

5-10. Functional elements of the shot process are the linkage between the Soldier, the weapon system, the environment, and the target that directly impact the shot process and ultimately the consistency, accuracy, and precision of the shot. When used appropriately, they build a greater understanding of any engagement.

5-11. The functional elements are interdependent. A accurate shot, regardless of weapon system, requires the Soldier to establish, maintain, and sustain —

- **Stability** – the Soldier stabilizes the weapon to provide a consistent base to fire from and maintain through the shot process until the recoil pulse has ceased. This process includes how the Soldier holds the weapon, uses structures or objects to provide stability, and the Soldier's posture on the ground during an engagement.
- **Aim** – the continuous process of orienting the weapon correctly, aligning the sights, aligning on the target, and the appropriate lead and elevation (hold) during a target engagement.
- **Control** – all the conscious actions of the Soldier before, during, and after the shot process that the Soldier specifically is in control of. The first of which is trigger control. This includes whether, when, and how to engage. It incorporates the Soldier as a function of safety, as well as the ultimate responsibility of firing the weapon.
- **Movement** – the process of the Soldier moving during the engagement process. It includes the Soldier's ability to move laterally, forward, diagonally, and in a retrograde manner while maintaining stabilization, appropriate aim, and control of the weapon.

5-12. These elements define the tactical engagement that require the Soldier to make adjustments to determine appropriate actions, and compensate for external influences on their shot process. When all elements are applied to the fullest extent, Soldiers will be able to rapidly engage targets with the highest level of precision.

5-13. Time, target size, target distance, and the Soldier's skills and capabilities determine the amount of effort required of each of the functional elements to minimize induced errors of the shot.

5-14. Each weapon, tactical situation, and sight system will have preferred techniques for each step in the shot process and within the functional elements to produce precision and accuracy in a timely manner. How fast or slow the shooter progresses through the process is based on target size, target distance, and shooter capability.

5-15. The most complex form of shooting is under combat conditions when the Soldier is moving, the enemy is moving, under limited visibility conditions. Soldiers and leaders must continue to refine skills and move training from the simplest shot to the most complex. Applying the functional elements during the shot process builds a firer's speed while maintaining consistency, accuracy, and precision during complex engagements.

5-16. Each of the functional elements and the Soldier actions to consider during the shot process are described later in this manual.

Chapter 5

TARGET ACQUISITION

5-17. Target acquisition is the ability of a Soldier to rapidly recognize threats to the friendly unit or formation. It is a critical Soldier function before any shot process begins. It includes the Soldier's ability to use all available optics, sensors, and information to detect potential threats as quickly as possible.

5-18. Target acquisition requires the Soldier to apply an acute attention to detail in a continuous process based on the tactical situation. The target acquisition process includes all the actions a Soldier must execute rapidly:
- **Detect** potential threats (target detection).
- **Identify** the threat as friend, foe, or noncombatant (target identification).
- **Prioritize** the threat(s) based on the level of danger they present (target prioritization).

TARGET DETECTION

5-19. Effective target detection requires a series of skills that Soldiers must master. Detection is an active process during combat operations with or without a clear or known threat presence. All engagements are enabled by the Soldier's detection skills, and are built upon three skill sets:
- **Scan and search** – a rapid sequence of various techniques to identify potential threats. Soldier scanning skills determine potential areas where threats are most likely to appear.
- **Acquire** – a refinement of the initial scan and search, based on irregularities in the environment.
- **Locate** – the ability to determine the general location of a threat to engage with accuracy or inform the small unit leader of contact with a potential threat.

Scan and Search

5-20. Scanning and searching is the art of observing an assigned sector. The goal of the scan and search is a deliberate detection of potential threats based on irregularities in the surrounding environment. This includes irregular shapes, colors, heat sources, movement, or actions the Soldier perceives as being "out of place," as compared to the surrounding area.

5-21. Soldiers use five basic search and scan techniques to detect potential threats in combat situations:
- **Rapid scan** – used to detect obvious signs of threat activity quickly. It is usually the first method used, whether on the offense or fighting in the defense.
- **Slow scan** – if no threats are detected during the rapid scan, Soldiers conduct the more deliberate scan using various optics, aiming devices, or sensors. The slow scan is best conducted in the defense or during slow movement or tactical halts.

- **Horizontal scan** – are used when operating in restricted or urban terrain. It is a horizontal sweeping scan that focuses on key areas where potential threats may be over watching their movement or position.
- **Vertical scan** – an up and down scan in restricted or urban environments to identify potential threats that may be observing the unit from an elevated position.
- **Detailed search** – used when no threats are detected using other scanning methods. The detailed search uses aiming devices, thermal weapon systems, magnified optics, or other sensors to slowly and methodically review locations of interest where the Soldier would be positioned if they were the threat (where would I be if I were them?)

Acquire

5-22. Target acquisition is the discovery of any object in the operational environment such as personnel, vehicles, equipment, or objects of potential military significance. Target acquisition occurs during target scan and search as a direct result of observation and the detection process.

5-23. During the scan and search, Soldiers are looking for "target signatures," which are signs or evidence of a threat. Tactically, Soldiers will be looking for threat personnel, obstacles or mines (including possible improvised explosive devices [IEDs]), vehicles, or anti-tank missile systems. These target signatures can be identified with sight, sound, or smell.

Detection Best Practices

5-24. Threat detection is a critical skill that requires thoughtful application of the sensors, optics, and systems at the Soldier's disposal. Finding potential threats as quickly and effectively as possible provides the maximum amount of time to defeat the threat. Soldiers should be familiar with the following best practices to increase target detection:
- Scan with the unaided eye first, then with a magnified optic.
- Practice using I2 and thermal optics in tandem during limited visibility.
- Understand the difference between I2 and thermal optics; what they can "see" and what they can't. (See chapter 4 of this publication.)
- Thermal optics are the preferred sight for target acquisition and engagement, day or night.
- Don't search in the same area as others in the small unit. Overlap, but do not focus on the same sector.
- Practice extreme light discipline during limited visibility including IR light discipline.
- Think as the threat. Search in areas that would be most advantageous from their perspective.
- Detecting threats is exponentially more difficult when operating in a chemical, biological, radiological, nuclear (CBRN) environment. Practice detection skills with personal protective equipment (PPE) individual

protective equipment (IPE) and understand the increased constraints and limitations, day and night.

Locate

5-25. Target location is the determination of where a target is in your operational environment in relation to the shooter, small unit, or element. Locating a target or series of targets occurs as a result of the search and acquisition actions of each Soldier in the small unit.

5-26. Once a target is located, the threat location can be rapidly and efficiently communicated to the rest of the unit. Methods used to announce a located target depend on the individual's specific position, graphic control measures for the operation, unit SOP, and time available.

TARGET IDENTIFICATION

5-27. Identifying (or discriminating) a target as friend, foe, or noncombatant (neutral) is the second step in the target acquisition process. The Soldier must be able to positively identify the threat into one of three classifications:

- **Friend.** Any force, U.S. or allied, that is jointly engaged in combat operations with an enemy in a theater of operation.
- **Foe (enemy combatant).** Any individual who has engaged acts against the U.S. or its coalition partners in violation of the laws and customs of war during an armed conflict.
- **Noncombatants.** Personnel, organizations, or agencies that are not taking a direct part in hostilities. This includes individuals such as medical personnel, chaplains, United Nations observers, or media representatives or those out of combat such as the wounded or sick. Organizations like the Red Cross or Red Crescent can be classified as noncombatants.

5-28. The identification process is complicated by the increasing likelihood of having to discriminate between friend/foe and combatant/noncombatant in urban settings or restricted terrain. To mitigate fratricide and unnecessary collateral damage, Soldiers use all of the situational understanding tools available and develop tactics, techniques, and procedures for performing target discrimination.

Fratricide Prevention

5-29. Units have other means of designating friendly vehicles from the enemy. Typically, these marking systems are derived from the unit tactical standard operating procedure (TACSOP) or other standardization publications, and applied to the personnel, small units, or vehicles as required:

- **Markings.** Unit markings are defined within the unit SOP. They distinctly identify a vehicle as friendly in a standardized manner.
- **Panels.** VS-17 panels provide a bright recognition feature that allows Soldiers to identify friendly vehicles through the day sight during unlimited visibility. Panels do not provide a thermal signature.

Employment

- **Lighting.** Chemical or light emitting diode lights provide a means of marking vehicles at night. However, chemical lights are not visible through a thermal sight. An IR variant is available for use with night vision devices. Lighting systems do not provide for thermal identification during day or limited visibility operations.
- **Beacons and Strobes.** Beacons and strobes are unit-procured, small-scale, compact, battery-operated flashing devices that operate in the near infrared wavelength. They are clearly visibly through night vision optics, but cannot be viewed through thermal optics.

   Note. Beacons and strobes generate illumination signals that can only be viewed by I2 optics. The signal *cannot be viewed* by thermal optics. Leaders and Soldiers are required to be aware of which optic can effectively view these systems when developing their SOPs and when using them in training or combat.

   Beacons and strobes have the potential to be viewed by enemy elements with night vision capabilities. Units should tailor use of the beacon based on METT-TC.

- **Symbols.** Unit symbols may be used to mark friendly vehicles. An inverted V, for example, painted on the flanks, rear, and fronts of a vehicle, aid in identifying a target as friendly. These are typically applied in an area of operations and not during training. Symbol marking systems do not provide for thermal identification during day or limited visibility operations.

TARGET PRIORITIZATION

5-30. When faced with multiple targets, the Soldier must prioritize each target and carefully plan his shots to ensure successful target engagement. Mental preparedness and the ability to make split-second decisions are the keys to a successful engagement of multiple targets. The proper mindset will allow the Soldier to react instinctively and control the pace of the battle, rather than reacting to the adversary threat.

5-31. Targets are prioritized into three threat levels—
- **Most dangerous.** A threat that has the capability to defeat the friendly force and is preparing to do so. These targets must be defeated immediately.
- **Dangerous.** A threat that has the capability to defeat the friendly force, but is not prepared to do so. These targets are defeated after all most dangerous targets are eliminated.
- **Least dangerous.** Any threat that does not have the ability to defeat the friendly force, but has the ability to coordinate with other threats that are more prepared. These targets are defeated after all threats of a higher threat level are defeated.

5-32. When multiple targets of the same threat level are encountered, the targets are prioritized according to the threat they represent. The standard prioritization of targets establishes the order of engagement. Firers engage similar threats by the following guide:
- **Near before far.**
- **Frontal before flank.**
- **Stationary before moving.**

5-33. The prioritization of targets provides a control mechanism for the shooter, and facilitates maintaining overmatch over the presented threats. Firers should be prepared deviate from the prioritization guide based on the situation, collective fire command, or changes to the target's activities.

Chapter 6
Stability

Stability is the ability of the Soldier to create a stable firing platform for the engagement. The Soldier stabilizes the weapon to provide a consistent base from which to fire from and maintain through the shot process until the recoil impulse has ceased. This process includes how the Soldier holds the weapon, uses structures or objects to provide stability, and the Soldier's posture on the ground during an engagement. A stable firing platform is essential during the shot process, whether the Soldier is stationary or moving.

This chapter provides the principles of developing a stable firing platform, describes the interaction between the Soldier, weapon, the surroundings, and the methods to achieve the greatest amount of stability in various positions. It explains how the stability functional element supports the shot process and interacts and integrates the other three elements. Stability provides a window of opportunity to maintain sight alignment and sight picture for the most accurate shot.

SUPPORT

6-1. Stability is provided through four functions: support, muscle relaxation, natural point of aim, and recoil management. These functions provide the Soldier the means to best stabilize their weapon system during the engagement process.

6-2. The placement or arrangement of sandbags, equipment, or structures that directly provide support to the upper receiver of the weapon to provide increased stability. This includes the use of a bipod or vertical foregrip, bone and muscle support provided by the shooter to stabilize the rifle.

6-3. Support can be natural or artificial or a combination of both. Natural support comes from a combination of the shooter's bones and muscles. Artificial support comes from objects outside the shooter's body. The more support a particular position provides, the more stable the weapon.

- **Leg Position.** The position of the legs varies greatly depending on the firing position used. The position may require the legs to support the weight of the Soldier's body, support the firing elbow, or to meet other requirements for the firing position. When standing unsupported, the body is upright with the legs staggered and knees slightly bent. In the prone, the firer's legs may be spread apart flat on the ground or bent at the knee. In the sitting position, the legs may also serve an intricate part of the firing position.

Chapter 6

- **Stance/Center of Gravity.** The physical position of a Soldier before, during, and after the shot that relates to the firer's balance and posture. The position/center of gravity does not apply when firing from the prone position. The position/center of gravity specifically relates to the Soldier's ability to maintain the stable firing platform during firing, absorbing the recoil impulses, and the ability to aggressively lean toward the target area during the shot process.
- **Firing Elbow.** The placement of the firing elbow during the shot process. Proper elbow placement provides consistent firing hand grip while standing, sitting, or kneeling, and provides support stability in the prone position.
- **Nonfiring Elbow.** The Soldier's placement of the nonfiring elbow during the shot process supports the rifle in the all positions.
- **Firing Hand.** Proper placement of the firing hand will aid in trigger control. Place the pistol grip in the 'V' formed between the thumb and index finger. The pressure applied is similar to a firm handshake grip. Different Soldiers have different size hands and lengths of fingers, so there is no set position of the finger on the trigger. To grip the weapon, the Soldier places the back strap of the weapon's pistol grip high in the web of his firing side hand between his thumb and index (trigger) finger. The Soldier's trigger finger is indexed on the lower receiver, well outside the trigger guard and off the magazine release to prevent inadvertent release of the magazine. The firing hand thumb (or trigger finger for left-handed firers) is indexed on top of the safety selector switch. The Soldier grasps the pistol grip with his remaining three fingers ensuring there is no gap between his middle finger and the trigger guard.
- **Nonfiring Hand.** Proper placement of the non-firing hand is based on the firing position and placement of the non-firing elbow to provide the stability of the weapon. Placement is adjusted during supported and unsupported firing to maximize stability. The non-firing hand is placed as far forward as comfortable without compromising the other elements of the position or inducing extreme shooter-gun angle.
    - The nonfiring hand supports the weight of the rifle by grasping the fore arm. It should be a firm but relaxed grip. In all positions it should be as close to the handguard as naturally possible to aid in recoil management.
    - If possible, the firer should strive to have the thumb of the nonfiring hand provide downward force on the handguard. The pressure will provide the necessary force to assist in the management of the muzzle rise from recoil.
    - In all positions it should be as close to the end of the handguard as naturally possible to aid in recoil management.
    - Due to limited space on current MWS rails the above may not be possible but consideration should be given while mounting lasers to achieve an extended grip.
- **Butt Stock.** Correct placement of the butt stock in the firing shoulder will aid in achieving a solid stock weld. Side to side placement will vary

Stability

depending on equipment worn while firing. The butt stock is placed high enough in the shoulder to allow for an upright head position.

- The vertical placement of the butt stock will vary from firing position to firing position. A general guideline to follow is: the higher the position from the ground, the higher the butt stock will be in the shoulder.
- The term "butt stock" refers to both the butt stock (M16-series) and collapsible butt stock (M4-series) for clarity.

- **Stock Weld.** Stock weld is the placement of the firer's head on the stock of the weapon. Correct stock weld is critical to sight alignment. The firer rests the full weight of the head on the stock. The head position is as upright as possible to give the best vision through the aiming device. It allows for scanning additional targets not seen through the aiming device.
    - When establishing the stock weld, bring the rifle up to your head, not your head down to the rifle. The firer's head will remain in the same location on the stock while firing, but the location may change when positions are changed. The bony portion of the cheek placed on the stock is the basic starting point. Soldiers adapt to their facial structure to find the optimal placement that allows for both sight alignment and repetitive placement.
    - Figure 6-1 shows the differences in head placement, which effects sight alignment. The firer on the right is NOT resting the full weight of their head on the stock. The picture on the left shows the skin of the firer's head being pushed down by the full weight of their head. This technique can be quickly observed and corrected by a peer coach.

Note. Soldiers' bodies vary with the amount of flesh and the bone structure of the face. Firers who apply downward force simply to achieve the appearance in the correct (left) image in figure 6-1, on page 6-4, will not have relaxation and will not have a repeatable placement. The goal is to have alignment with consistent placement.

Chapter 6

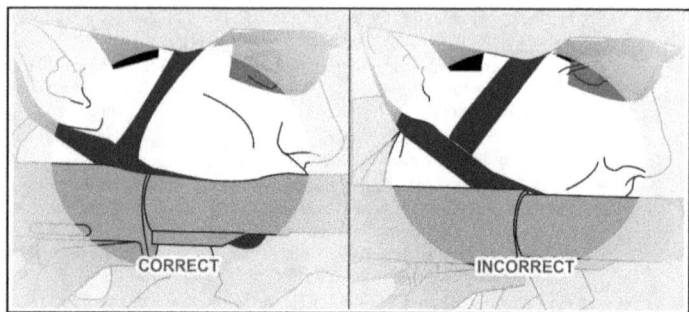

Figure 6-1. Stock weld

MUSCLE RELAXATION

6-4. Muscle relaxation is the ability of the Soldier to maintain orientation of the weapon appropriately during the shot process while keeping the major muscle groups from straining to maintain the weapon system's position. Relaxed muscles contribute to stability provided by support.

- Strained or fatigued muscles detract from stability.
- As a rule, the more support from the shooter's bones the less he requires from his muscles.
- The more skeletal support, the more stable the position, as bones do not fatigue or strain.
- As a rule, the less muscle support required, the longer the shooter can stay in position.

NATURAL POINT OF AIM

6-5. The natural point of aim is the point where the barrel naturally orients when the shooter's muscles are relaxed and support is achieved. The natural point of aim is built upon the following principles:

- The closer the natural point of aim is to the target, the less muscle support required.
- The more stable the position, the more resistant to recoil it is.
- More of the shooter's body on the ground equals a more stable position.
- More of the shooter's body on the ground equals less mobility for the shooter.

6-6. When a Soldier aims at a target, the lack of stability creates a wobble area, where the sights oscillate slightly around and through the point of aim. If the wobble area is larger than the target, the Soldier requires a steadier position or a refinement to their position to decrease the size of his wobble area before trigger squeeze.

Stability

Note. The steadier the position, the smaller the wobble area. The smaller the wobble area, the more precise the shot.

6-7. To check a shooter's natural point of aim, the Soldier should assume a good steady position and get to the natural pause. Close their eyes, go through one cycle, and then open their eyes on the natural pause. Where the sights are laying at this time, is the natural point of aim for that position. If it is not on their point of aim for their target, they should make small adjustments to their position to get the reticle or front sight post back on their point of aim. The Soldier will repeat this process until the natural point of aim is on the point of aim on their target.

RECOIL MANAGEMENT

6-8. Recoil management is the result of a Soldier assuming and maintaining a stable firing position which mitigates the disturbance of one's sight picture during the cycle of function of the weapon.

6-9. The Soldier's firing position manages recoil using support of the weapon system, the weight of their body, and the placement of the weapon during the shot process. Proper recoil management allows the sights to rapidly return to the target and allows for faster follow up shots.

SHOOTER–GUN ANGLE

6-10. The shooter gun-angle is the relationship between the shooters upper body and the direction of the weapon. This angle is typically different from firing position to firing position, and directly relates to the Soldier's ability to control recoil. Significant changes in the shooter-gun angle can result in eye relief and stock weld changes.

Note. Units with a mix of left and right handed shooters can take advantage of each Soldiers' natural carry positions, and place left-handed shooters on the right flanks, and right-handed shooters on the left flanks, as their natural carry alignment places the muzzle away from the core element, and outward toward potential threats, and reduces the challenges of firing when moving laterally.

FIELD OF VIEW

6-11. The field of view is the extent that the human eye can see at any given moment. The field of view is based on the Soldier's view ***without*** using magnification, optics, or thermal devices. The field of view is what the Soldier sees, and includes the areas where the Soldier can detect potential threats.

CARRY POSITIONS

6-12. There are six primary carry positions. These positions may be directed by the leader, or assumed by the Soldier based on the tactical situation. The primary positions are—
- Hang.
- Safe hang.
- Collapsed low ready.
- Low ready.
- High ready.
- Ready (or ready-up).

Stability

HANG

6-13. Soldiers use the hang when they need their hands for other tasks and no threat is present or likely (see figure 6-2). The weapon is slung and the safety is engaged. The hang carry should not be used when the weapon control status is RED. The reduced security of the weapon may cause the mechanical safety select lever to unintentionally move to SEMI or BURST/AUTO.

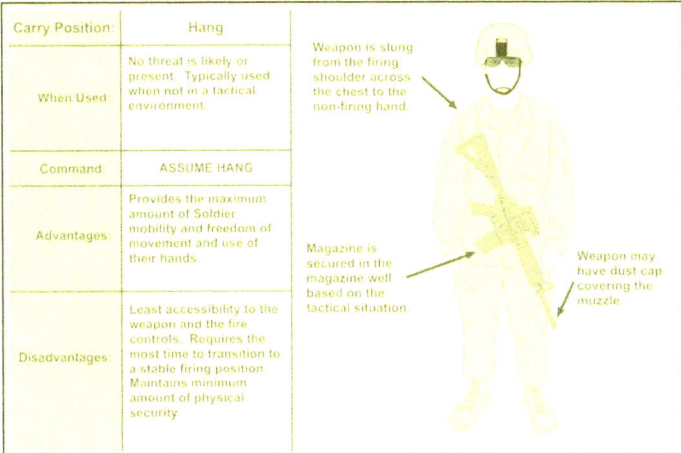

Carry Position:	Hang
When Used:	No threat is likely or present. Typically used when not in a tactical environment.
Command:	ASSUME HANG
Advantages:	Provides the maximum amount of Soldier mobility and freedom of movement and use of their hands.
Disadvantages:	Least accessibility to the weapon and the fire controls. Requires the most time to transition to a stable firing position. Maintains minimum amount of physical security.

Weapon is slung from the firing shoulder across the chest to the non-firing hand.

Magazine is secured in the magazine well based on the tactical situation.

Weapon may have dust cap covering the muzzle.

Figure 6-2. Hang carry example

Chapter 6

SAFE HANG

6-14. The safe hang is used when no immediate threat is present and the hands are not necessary (see figure 6-3). In the safe hang carry, the weapon is slung, the safety is engaged, and the Soldier has gripped the rifle's pistol grip. The Soldier sustains Rule 3, keeping the finger off the trigger until ready to engage when transitioning to the ready or ready up position.

6-15. In this position, the Soldier can move in any direction while simultaneously maintaining his muzzle oriented at the ground by using his firing hand. This carry provides control of the weapon, flexibility in movement, and positive control of the weapon's fire controls.

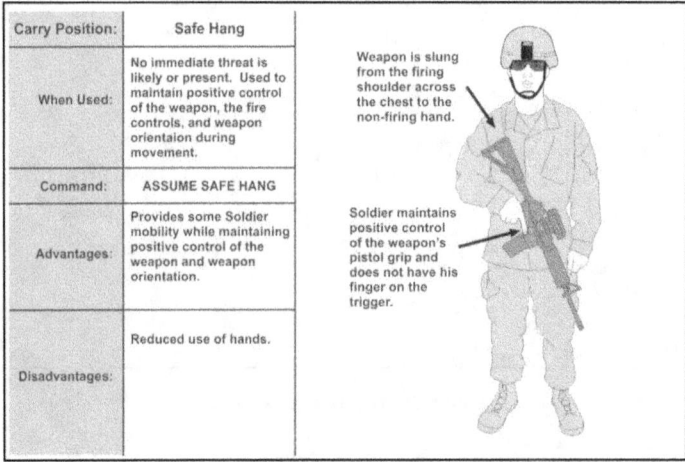

Figure 6-3. Safe hang example

Stability

COLLAPSED LOW READY

6-16. The collapsed low ready is used when a greater degree of muzzle control and readiness to respond to threats or weapon retention is necessary (such as crowded environments). In the collapsed low ready, the firing hand is secure on the weapon's pistol grip. The non-firing hand is placed on the hand guards or vertical foregrip (see figure 6-4).

6-17. This carry allows a Soldier to navigate crowded or restrictive environments while simultaneously minimizing or eliminating his muzzle covering (flagging) by maintaining positive control of the muzzle orientation.

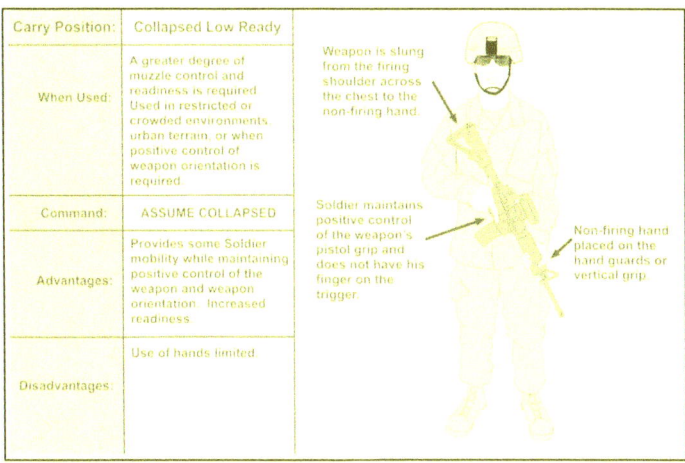

Figure 6-4. Collapsed low ready example

Chapter 6

LOW READY

6-18. The low ready provides the highest level of readiness and with the maximum amount of observable area for target acquisition purposes

6-19. In the low ready position, the weapon is slung, the butt stock is in the Soldier's shoulder, and the muzzle is angled down at a 30- to 45-degree angle and oriented towards the Soldier's sector of fire.

6-20. Firing hand is positioned on the pistol grip with the index finger straight and out of the trigger guard. The thumb is placed on the selector lever with the lever placed on safe. From this carry, the Soldier is ready to engage threats within a very short amount of time with minimal movement. (See figure 6-5).

6-21. Observation is maintained to the sector of fire. The Soldier looks over the top of his optics or sights to maintain situation awareness of his sector. The Soldier's head remains upright.

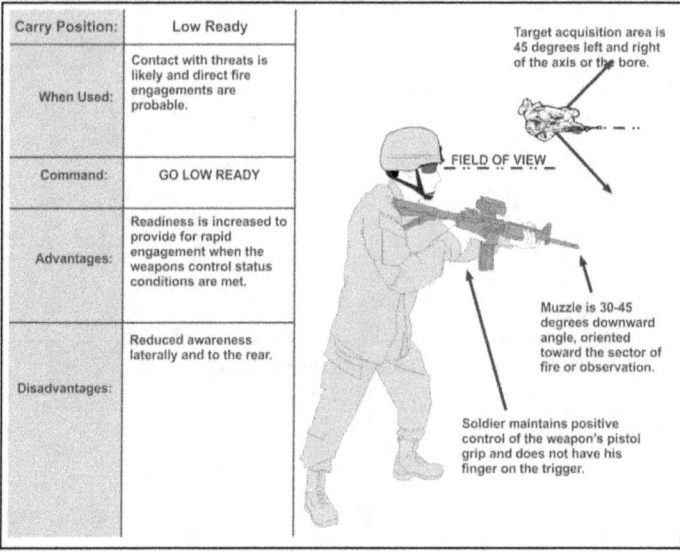

Figure 6-5. Low ready position

Stability

HIGH READY

6-22. The high ready is used when the Soldier's sector of fire includes areas overhead or when an elevated muzzle orientation is appropriate for safety (see figure 6-6). The high ready carry is used when contact is likely.

6-23. In the high ready, the weapon is slung, butt stock is in the armpit, the muzzle angled up to at least a 45-degree angle and oriented toward the Soldier's sector of fire—ensuring no other Soldiers are flagged.

6-24. The firing hand remains in the same position as the low ready. The non-firing side hand can be free as the weapon is supported by the firing side hand and armpit.

6-25. This position is not as effective as the low ready for several reasons: it impedes the field of view, flags friendlies above the sector of fire, and typically takes longer to acquire the target.

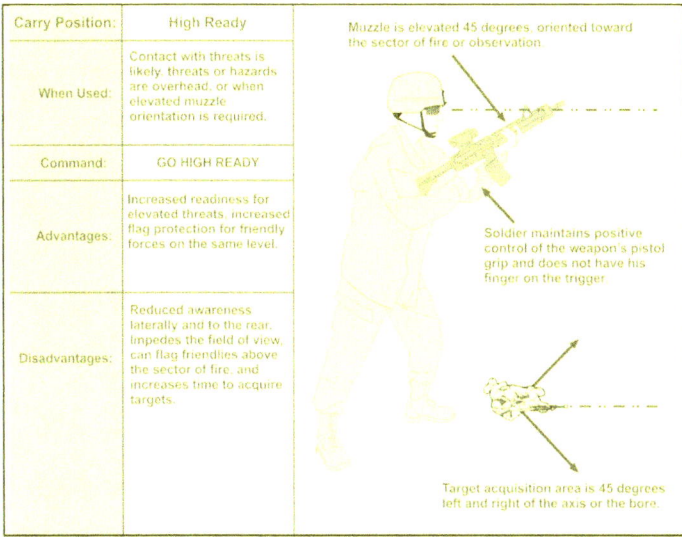

Figure 6-6. High ready position

Chapter 6

READY OR READY-UP

6-26. The ready is used when enemy contact is imminent (see figure 6-7). This carry is used when the Soldier is preparing or prepared to engage a threat.

6-27. In the ready, the weapon is slung, the toe of the butt stock is in the Soldier's shoulder, and muzzle is oriented toward a threat or most likely direction of enemy contact. The Soldier is looking through his optics or sights. His non-firing side hand remains on the hand guards or the vertical foregrip.

6-28. The firing hand remains on the pistol grip with the firing finger off the trigger until the decision to engage a target is made.

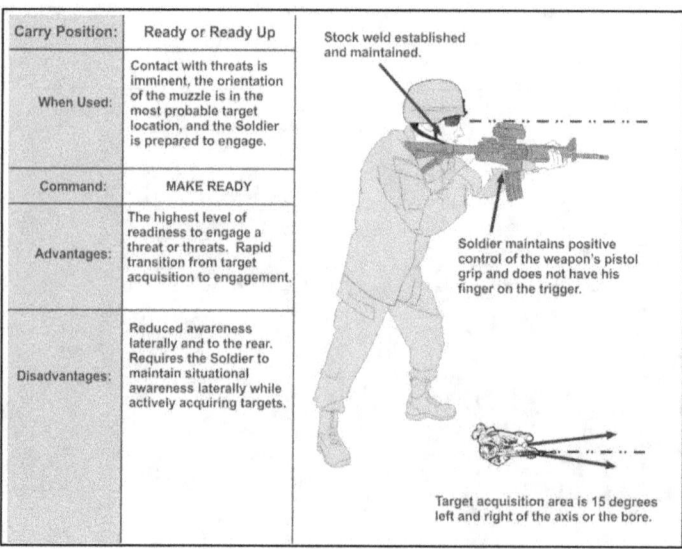

Carry Position:	Ready or Ready Up
When Used:	Contact with threats is imminent, the orientation of the muzzle is in the most probable target location, and the Soldier is prepared to engage.
Command:	MAKE READY
Advantages:	The highest level of readiness to engage a threat or threats. Rapid transition from target acquisition to engagement.
Disadvantages:	Reduced awareness laterally and to the rear. Requires the Soldier to maintain situational awareness laterally while actively acquiring targets.

Stock weld established and maintained.

Soldier maintains positive control of the weapon's pistol grip and does not have his finger on the trigger.

Target acquisition area is 15 degrees left and right of the axis or the bore.

Figure 6-7. Ready position or up position

STABILIZED FIRING

6-29. The Soldier must stabilize their weapon, whether firing from a stationary position or while on the move. To create a stabilized platform, Soldiers must understand the physical relationship between the weapon system, the shooter's body, the ground, and any other objects touching the weapon or shooter's body. The more contact the shooter has to the ground will determine how stable and effective the position is. The situation and tactics will determine the actual position used.

6-30. When a shooter assumes a stable firing position, movement from muscle tension, breathing, and other natural activities within the body will be transferred to the weapon and must be compensated for by the shooter.

6-31. Failing to create an effective platform to fire from is termed a *stabilization failure*. A stabilization failure occurs when a Soldier fails to:

- Control the movement of the barrel during the arc of movement
- Adequately support the weapon system
- Achieve their natural point of aim.

6-32. These failures compound the firing occasion's errors, which directly correlate to the accuracy of the shot taken. To maximize the Soldier's stability during the shot process, they correctly assume various firing positions when stationary, or offset the induced errors with other firing skills during tactical movement.

6-33. As a rule, positions that are lower to the ground provide a higher level of stability. When the center of gravity elevates the level of stability decreases as shown in figure 6-8.

Chapter 6

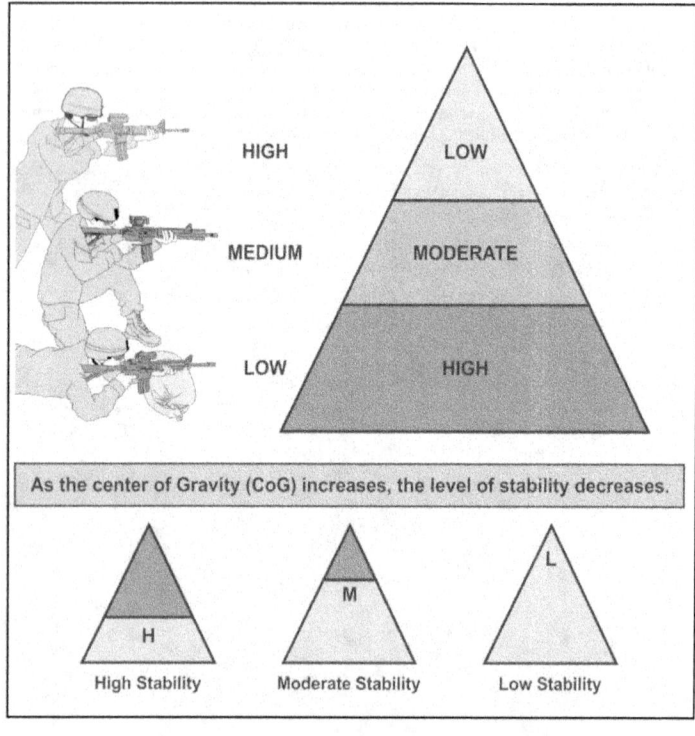

Figure 6-8. Firing position stability example

Stability

FIRING POSITIONS

6-34. The nature of combat will not always allow time for a Soldier to get into a particular position. Soldiers need to practice firing in a variety of positions, including appropriate variations. There are 12 firing positions with variations that are common to all Soldiers. The positions are listed highest to lowest. The primary position is listed in bold, with the position variations in italics:

- **Standing**
  - *Standing, unsupported.*
  - *Standing, supported.*
- **Squatting** – This position allows for rapid engagement of targets when an obstruction blocks the firer from using standard positions. It provides the firer a fairly well supported position by simply squatting down to engage, then returning to a standing position once the engagement is complete. The squatting position is generally unsupported.
- **Kneeling** – The kneeling position is very common and useful in most combat situations. The kneeling position can be supported or unsupported.
  - *Kneeling, unsupported.*
  - *Kneeling, supported.*
- **Sitting** – There are three types of sitting positions: crossed-ankle, crossed-leg, and open-leg. All positions are easy to assume, present a medium silhouette, provide some body contact with the ground, and form a stable firing position. These positions allow easy access to the sights for zeroing.
  - *Sitting, crossed ankle.*
  - *Sitting, crossed leg.*
  - *Sitting, open leg.*
- **Prone** – The prone position is the most stable firing position due to the amount of the Soldier's body is in contact with the ground. The majority of the firer's frame is behind the rifle to assist with recoil management.
  - *Prone, unsupported.*
  - *Prone, supported.*
  - *Prone, roll-over.*
  - *Prone, reverse roll-over.*

6-35. Soldiers must practice the positions dry frequently to establish their natural point of aim for each position, and develop an understanding of the restrictive nature of their equipment during execution. With each dry repetition, the Soldier's ability to change positions rapidly and correctly are developed, translating into efficient movement and consistent stable firing positions.

6-36. Each of these firing positions is described using in a standard format using the terms defined earlier.

Chapter 6

STANDING, UNSUPPORTED

6-37. This position should be used for closer targets or when time is not available to assume a steadier position such as short range employment. The upper body should be leaned slightly forward to aid in recoil management. The key focus areas for the standing supported position are applied as described in figure 6-9 below:

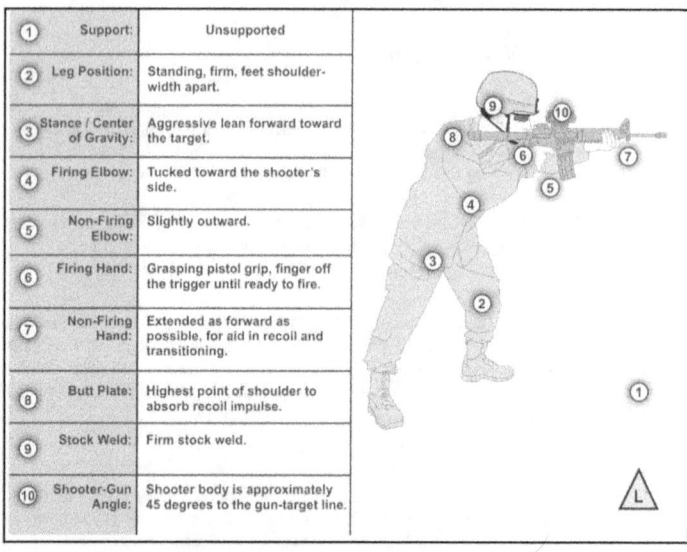

①	Support:	Unsupported
②	Leg Position:	Standing, firm, feet shoulder-width apart.
③	Stance / Center of Gravity:	Aggressive lean forward toward the target.
④	Firing Elbow:	Tucked toward the shooter's side.
⑤	Non-Firing Elbow:	Slightly outward.
⑥	Firing Hand:	Grasping pistol grip, finger off the trigger until ready to fire.
⑦	Non-Firing Hand:	Extended as forward as possible, for aid in recoil and transitioning.
⑧	Butt Plate:	Highest point of shoulder to absorb recoil impulse.
⑨	Stock Weld:	Firm stock weld.
⑩	Shooter-Gun Angle:	Shooter body is approximately 45 degrees to the gun-target line.

Figure 6-9. Standing, unsupported example

Stability

STANDING, SUPPORTED

6-38. Soldier should ensure it is the handguard of the weapon NOT the barrel that is in contact with the artificial support. Barrels being in direct contact with artificial support will result in erratic shots. The standing supported position uses artificial support to steady the position (see figure 6-10.) Forward pressure should be applied by the rear leg and upper body to aid in recoil management. The key focus area for the standing supported position are applied in the following ways:

Nonfiring hand. The nonfiring hand will hold the hand guards firmly and push against the artificial support. Hand positioning will vary depending on the type of support used.

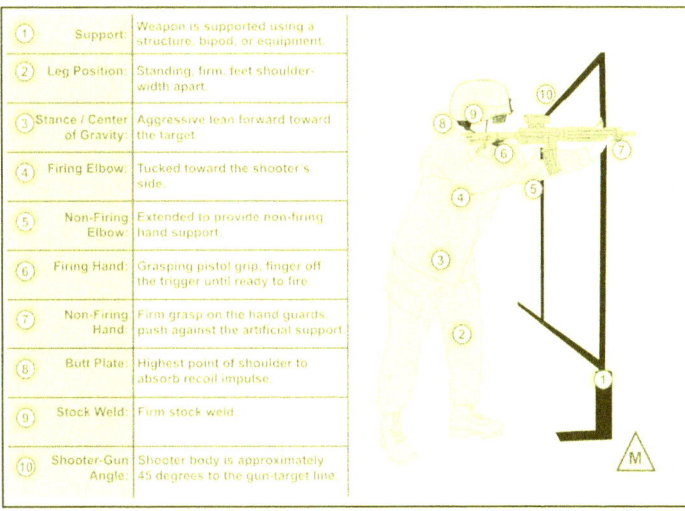

Figure 6-10. Standing, supported example

Chapter 6

SQUATTING

6-39. This position allows for rapid engagement of targets when an obstruction blocks the firer from using standard positions. It allows the firer a fairly stable position by simply squatting down to engage, then returning to a standing position after completing the engagement (see figure 6-11.)

6-40. Perform the following to assume a good squatting firing position:
- Face the target.
- Place the feet shoulder-width apart.
- Squat down as far as possible.
- Place the back of triceps on the knees ensuring there is no bone on bone contact.
- Place the firing hand on the pistol grip and the nonfiring hand on the upper hand guards.
- Place the weapon's butt stock high in the firer's shoulder pocket.

Note. The firer may opt to use pressure from firing hand to rotate weapon to place the magazine against the opposite forearm to aid in stabilization.

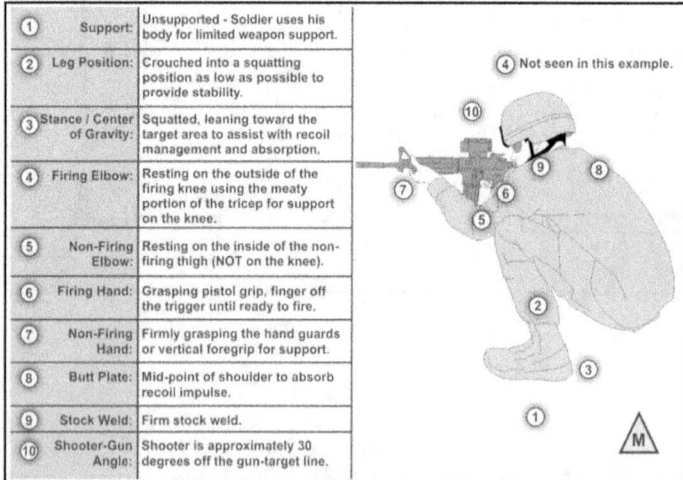

Figure 6-11. Squatting position

Stability

KNEELING, UNSUPPORTED

6-41. The kneeling unsupported position does not use artificial support. Figure 6-12 shows the optimum unsupported kneeling position. The firer should be leaning slightly forward into the position to allow for recoil management and quicker follow-up shots. The primary goal of this firing position is to establish the smallest wobble area possible. Key focus areas for kneeling, unsupported are:

- **Nonfiring elbow.** Place the non-firing elbow directly underneath the rifle as much as possible. The elbow should be placed either in front of or behind the kneecap. Placing the elbow directly on the kneecap will cause it to roll and increases the wobble area.
- **Leg position.** The non-firing leg should be bent approximately 90 degrees at the knee and be directly under the rifle. The firing-side leg should be perpendicular to the nonfiring leg. The firer may rest their body weight on the heel. Some firers lack the flexibility to do this and may have a gap between their buttocks and the heel.
- **Aggressive (stretch) kneeling.** All weight on non-firing foot, thigh to calf, upper body leaning forward, nonfiring triceps on non-firing knee, firing leg stretched behind for support. Highly effective for rapid fire and movement.

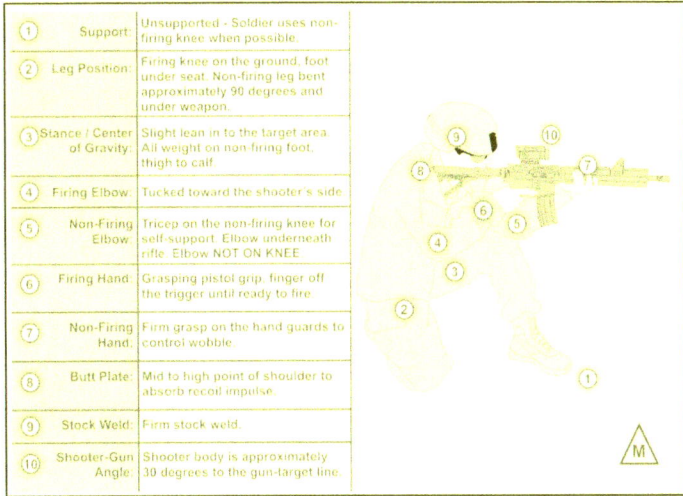

①	Support:	Unsupported - Soldier uses non-firing knee when possible.
②	Leg Position:	Firing knee on the ground, foot under seat. Non-firing leg bent approximately 90 degrees and under weapon.
③	Stance / Center of Gravity:	Slight lean in to the target area. All weight on non-firing foot, thigh to calf.
④	Firing Elbow:	Tucked toward the shooter's side.
⑤	Non-Firing Elbow:	Tricep on the non-firing knee for self-support. Elbow underneath rifle. Elbow NOT ON KNEE.
⑥	Firing Hand:	Grasping pistol grip, finger off the trigger until ready to fire.
⑦	Non-Firing Hand:	Firm grasp on the hand guards to control wobble.
⑧	Butt Plate:	Mid to high point of shoulder to absorb recoil impulse.
⑨	Stock Weld:	Firm stock weld.
⑩	Shooter-Gun Angle:	Shooter body is approximately 30 degrees to the gun-target line.

Figure 6-12. Kneeling, unsupported example

Chapter 6

KNEELING, SUPPORTED

6-42. The kneeling supported position uses artificial support to steady the position (see figure 6-13). Contact by the nonfiring hand and elbow with the artificial support is the primary difference between the kneeling supported and unsupported positions since it assists in the stability of the weapon. Body contact is good, but the barrel of the rifle must not touch the artificial support. Forward pressure is applied to aid in recoil management. The key focus areas for the kneeling supported position are applied in the following ways:

- **Nonfiring hand.** The nonfiring hand will hold the hand guards firmly and will also be pushed against the artificial support. Hand positioning will vary depending on the type of support used.
- **Nonfiring elbow.** The nonfiring elbow and forearm may be used to assist with the weapon's stability by pushing against the artificial support. The contact of the nonfiring elbow and forearm with the structure will vary depending on the support used and the angle to the target.

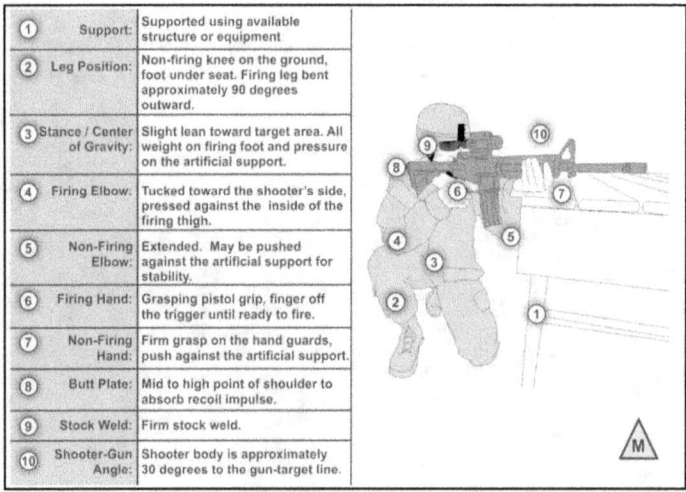

①	Support:	Supported using available structure or equipment
②	Leg Position:	Non-firing knee on the ground, foot under seat. Firing leg bent approximately 90 degrees outward.
③	Stance / Center of Gravity:	Slight lean toward target area. All weight on firing foot and pressure on the artificial support.
④	Firing Elbow:	Tucked toward the shooter's side, pressed against the inside of the firing thigh.
⑤	Non-Firing Elbow:	Extended. May be pushed against the artificial support for stability.
⑥	Firing Hand:	Grasping pistol grip, finger off the trigger until ready to fire.
⑦	Non-Firing Hand:	Firm grasp on the hand guards, push against the artificial support.
⑧	Butt Plate:	Mid to high point of shoulder to absorb recoil impulse.
⑨	Stock Weld:	Firm stock weld.
⑩	Shooter-Gun Angle:	Shooter body is approximately 30 degrees to the gun-target line.

Figure 6-13. Kneeling, supported example

Stability

SITTING, CROSSED-ANKLE

6-43. The sitting, crossed-ankle position provides a broad base of support and places most of the body weight behind the weapon (see figure 6-14). This allows quick shot recovery and recoil impulse absorption. Perform the following to assume a good crossed-ankle position:

- Face the target at a 10- to 30-degree angle.
- Place the nonfiring hand under the hand guard.
- Bend at knees and break fall with the firing hand.
- Push backward with feet to extend legs and place the buttocks to ground.
- Cross the non-firing ankle over the firing ankle.
- Bend forward at the waist.
- Place the non-firing elbow on the nonfiring leg below knee.
- Grasp the rifle butt with the firing hand and place into the firing shoulder pocket.
- Grasp the pistol grip with the firing hand.
- Lower the firing elbow to the inside of the firing knee.
- Place the cheek firmly against the stock to obtain a firm stock weld.
- Move the nonfiring hand to a location under the hand guard that provides the maximum bone support and stability for the weapon.

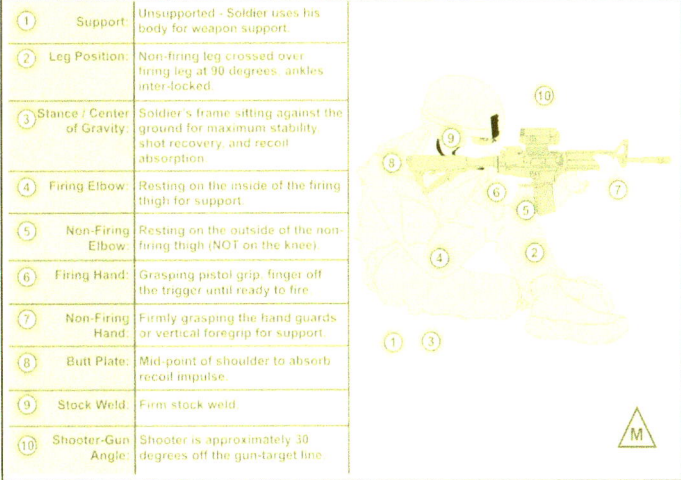

(1)	Support:	Unsupported - Soldier uses his body for weapon support.
(2)	Leg Position:	Non-firing leg crossed over firing leg at 90 degrees, ankles inter-locked.
(3)	Stance / Center of Gravity:	Soldier's frame sitting against the ground for maximum stability, shot recovery, and recoil absorption.
(4)	Firing Elbow:	Resting on the inside of the firing thigh for support.
(5)	Non-Firing Elbow:	Resting on the outside of the non-firing thigh (NOT on the knee).
(6)	Firing Hand:	Grasping pistol grip, finger off the trigger until ready to fire.
(7)	Non-Firing Hand:	Firmly grasping the hand guards or vertical foregrip for support.
(8)	Butt Plate:	Mid-point of shoulder to absorb recoil impulse.
(9)	Stock Weld:	Firm stock weld.
(10)	Shooter-Gun Angle:	Shooter is approximately 30 degrees off the gun-target line.

Figure 6-14. Sitting position—crossed ankle

SITTING, CROSSED-LEG

6-44. The crossed-leg sitting position provides a base of support and places most of the body weight behind the weapon for quick shot recovery (see figure 6-15). Soldiers may experience a strong pulse beat in this position due to restricted blood flow in the legs and abdomen. An increased pulse causes a larger wobble area.

6-45. Perform the following to assume a good crossed-leg position:
- Place the nonfiring hand under the hand guard.
- Cross the nonfiring leg over the firing leg.
- Bend at the knees and break the fall with the firing hand.
- Place the buttocks to the ground close to the crossed legs.
- Bend forward at the waist.
- Place the nonfiring elbow on the nonfiring leg at the bend of the knee.
- Establish solid butt stock position in the firing shoulder pocket.
- Grasp the pistol grip with the firing hand.
- Lower the firing elbow to the inside of the firing knee.
- Place the cheek firmly against the stock to obtain a firm stock weld.
- Place the non-firing hand under the hand guard to provide support.

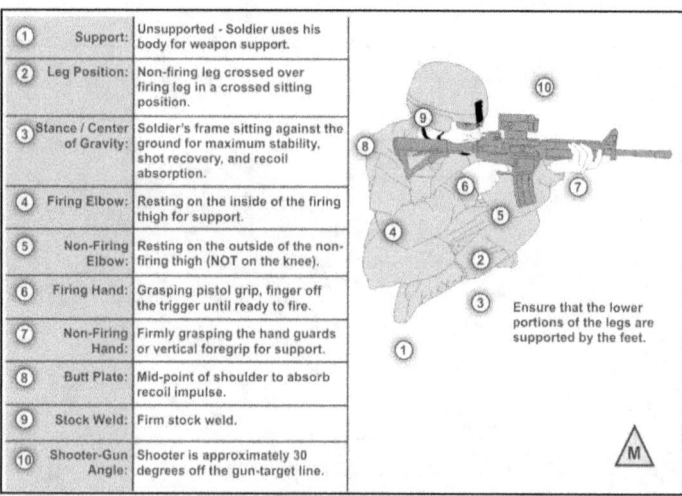

①	Support:	Unsupported - Soldier uses his body for weapon support.
②	Leg Position:	Non-firing leg crossed over firing leg in a crossed sitting position.
③	Stance / Center of Gravity:	Soldier's frame sitting against the ground for maximum stability, shot recovery, and recoil absorption.
④	Firing Elbow:	Resting on the inside of the firing thigh for support.
⑤	Non-Firing Elbow:	Resting on the outside of the non-firing thigh (NOT on the knee).
⑥	Firing Hand:	Grasping pistol grip, finger off the trigger until ready to fire.
⑦	Non-Firing Hand:	Firmly grasping the hand guards or vertical foregrip for support.
⑧	Butt Plate:	Mid-point of shoulder to absorb recoil impulse.
⑨	Stock Weld:	Firm stock weld.
⑩	Shooter-Gun Angle:	Shooter is approximately 30 degrees off the gun-target line.

Ensure that the lower portions of the legs are supported by the feet.

Figure 6-15. Sitting position—crossed-leg

Stability

SITTING, OPEN-LEG

6-46. The open-leg sitting position is the preferred sitting position when shooting with combat equipment (see figure 6-16). It places less of the body weight behind the weapon than the other sitting positions. Perform the following to assume a good open-leg position:

- Face the target at a 10 to 30 degree angle to the firing of the line of fire.
- Place the feet approximately shoulder width apart.
- Place the nonfiring hand under the hand guard.
- Bend at the knees while breaking the fall with the firing hand. Push backward with the feet to extend the legs and place the buttocks on ground.
- Place the both the firing and non-firing elbow inside the knees.
- Grasp the rifle butt with the firing hand and place into the firing shoulder pocket.
- Grasp the pistol grip with the firing hand.
- Lower the firing elbow to the inside of the firing knee.
- Place the cheek firmly against the stock to obtain a firm stock weld.
- Move nonfiring hand to a location under the hand guard that provides maximum bone support and stability for the weapon.

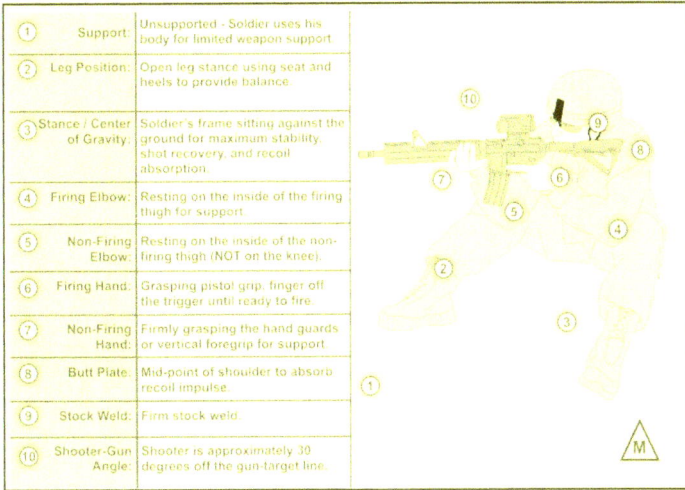

①	Support:	Unsupported - Soldier uses his body for limited weapon support.
②	Leg Position:	Open leg stance using seat and heels to provide balance.
③	Stance / Center of Gravity:	Soldier's frame sitting against the ground for maximum stability, shot recovery, and recoil absorption.
④	Firing Elbow:	Resting on the inside of the firing thigh for support.
⑤	Non-Firing Elbow:	Resting on the inside of the non-firing thigh (NOT on the knee).
⑥	Firing Hand:	Grasping pistol grip, finger off the trigger until ready to fire.
⑦	Non-Firing Hand:	Firmly grasping the hand guards or vertical foregrip for support.
⑧	Butt Plate:	Mid-point of shoulder to absorb recoil impulse.
⑨	Stock Weld:	Firm stock weld.
⑩	Shooter-Gun Angle:	Shooter is approximately 30 degrees off the gun-target line.

Figure 6-16. Sitting position—open leg

Chapter 6

PRONE, UNSUPPORTED

6-47. The prone unsupported position is not as stable as the prone supported position (see figure 6-17). Soldiers must build a stable, consistent position that focuses on the following key areas:

- **Firing hand.** The firer should have a firm handshake grip on the pistol grip and place their finger on the trigger where it naturally falls.
- **Nonfiring hand.** The nonfiring hand is placed to control the weapon and is comfortable.
- **Leg position.** The firer's legs may be either spread with heels as flat as possible on ground or the firing side leg may be bent at the knee to relieve pressure on the stomach.

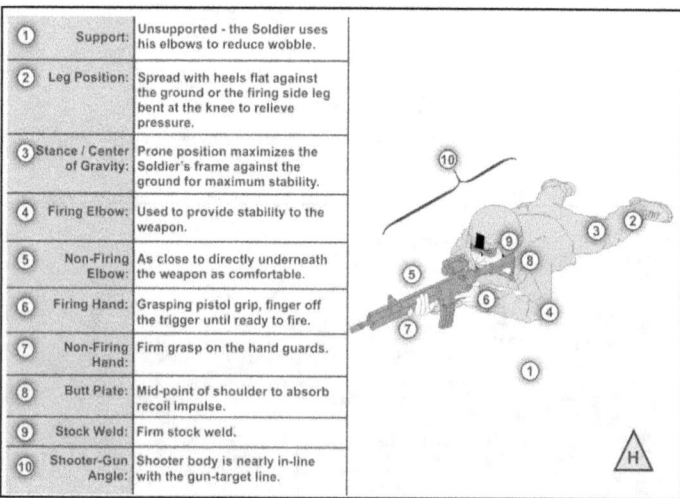

①	Support:	Unsupported - the Soldier uses his elbows to reduce wobble.
②	Leg Position:	Spread with heels flat against the ground or the firing side leg bent at the knee to relieve pressure.
③	Stance / Center of Gravity:	Prone position maximizes the Soldier's frame against the ground for maximum stability.
④	Firing Elbow:	Used to provide stability to the weapon.
⑤	Non-Firing Elbow:	As close to directly underneath the weapon as comfortable.
⑥	Firing Hand:	Grasping pistol grip, finger off the trigger until ready to fire.
⑦	Non-Firing Hand:	Firm grasp on the hand guards.
⑧	Butt Plate:	Mid-point of shoulder to absorb recoil impulse.
⑨	Stock Weld:	Firm stock weld.
⑩	Shooter-Gun Angle:	Shooter body is nearly in-line with the gun-target line.

Figure 6-17. Prone, unsupported example

Note. The magazine can be rested on the ground while using the prone unsupported position. Firing with the magazine on the ground will NOT induce a malfunction.

PRONE, SUPPORTED

6-48. The prone supported position allows for the use of support, such as sandbags (see figure 6-18). Soldiers must build a stable, consistent position that focuses on the following key areas:

- **Firing hand.** The firer should have a firm handshake grip on the pistol grip and place their finger on the trigger where it naturally falls.
- **Nonfiring hand.** The nonfiring hand is placed to maximize control the weapon and where it is comfortable on the artificial support.
- **Leg position.** The firer's legs may be either spread with heels as flat as possible on ground or the firing side leg may be bent at the knee to relieve pressure on the stomach.
- **Artificial support.** The artificial support should be at a height that allows for stability without interfering with the other elements of the position.

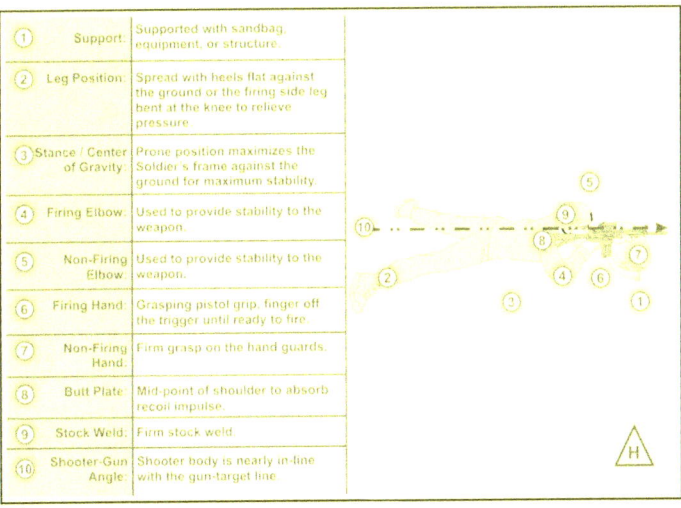

①	Support:	Supported with sandbag, equipment, or structure.
②	Leg Position:	Spread with heels flat against the ground or the firing side leg bent at the knee to relieve pressure.
③	Stance / Center of Gravity	Prone position maximizes the Soldier's frame against the ground for maximum stability.
④	Firing Elbow:	Used to provide stability to the weapon.
⑤	Non-Firing Elbow:	Used to provide stability to the weapon.
⑥	Firing Hand:	Grasping pistol grip, finger off the trigger until ready to fire.
⑦	Non-Firing Hand:	Firm grasp on the hand guards.
⑧	Butt Plate:	Mid-point of shoulder to absorb recoil impulse.
⑨	Stock Weld:	Firm stock weld.
⑩	Shooter-Gun Angle:	Shooter body is nearly in-line with the gun-target line.

Figure 6-18. Prone, supported example

Chapter 6

PRONE, ROLL-OVER

6-49. This position allows the firer to shoot under obstacles or cover that would not normally be attainable from the standard conventional prone position (see figure 6-19). With this position, the bullet trajectory will be off compared to the line of sight and increase with distance from the firer.

> For example, in the figure below the sights are rotated to the right. The trajectory of the bullet will be lower than and to the right of point of aim. This error will increase with range.

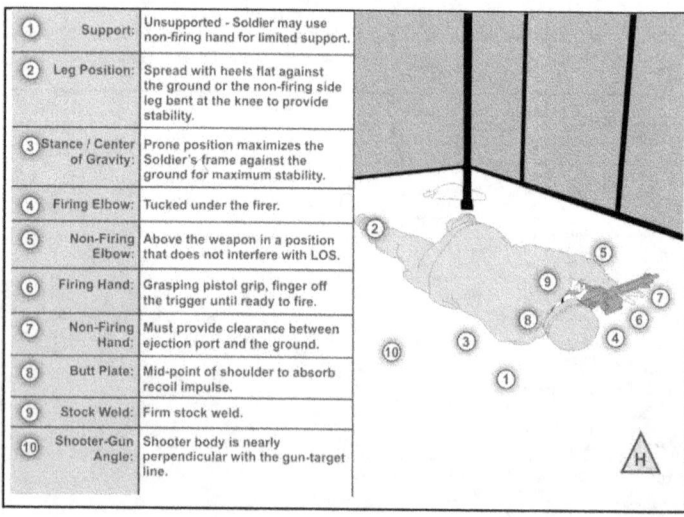

①	Support:	Unsupported - Soldier may use non-firing hand for limited support.
②	Leg Position:	Spread with heels flat against the ground or the non-firing side leg bent at the knee to provide stability.
③	Stance / Center of Gravity:	Prone position maximizes the Soldier's frame against the ground for maximum stability.
④	Firing Elbow:	Tucked under the firer.
⑤	Non-Firing Elbow:	Above the weapon in a position that does not interfere with LOS.
⑥	Firing Hand:	Grasping pistol grip, finger off the trigger until ready to fire.
⑦	Non-Firing Hand:	Must provide clearance between ejection port and the ground.
⑧	Butt Plate:	Mid-point of shoulder to absorb recoil impulse.
⑨	Stock Weld:	Firm stock weld.
⑩	Shooter-Gun Angle:	Shooter body is nearly perpendicular with the gun-target line.

Figure 6-19. Prone, roll-over example

Stability

PRONE, REVERSE ROLL-OVER

6-50. This position is primarily used when the firer needs to keep behind cover that is too low to use while in a traditional prone position (see figure 6-20). The bullet's trajectory will be off considerably at long distances while in this position.

6-51. This position is the most effective way to support the weapon when the traditional prone is too low to be effective and where a kneeling position is too high to gain cover or a solid base for support.

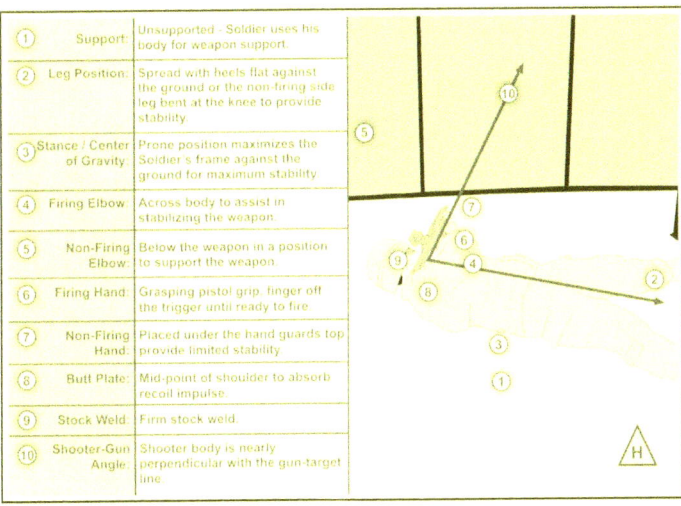

①	Support:	Unsupported - Soldier uses his body for weapon support.
②	Leg Position:	Spread with heels flat against the ground or the non-firing side leg bent at the knee to provide stability.
③	Stance / Center of Gravity:	Prone position maximizes the Soldier's frame against the ground for maximum stability.
④	Firing Elbow:	Across body to assist in stabilizing the weapon.
⑤	Non-Firing Elbow:	Below the weapon in a position to support the weapon.
⑥	Firing Hand:	Grasping pistol grip, finger off the trigger until ready to fire.
⑦	Non-Firing Hand:	Placed under the hand guards to provide limited stability.
⑧	Butt Plate:	Mid-point of shoulder to absorb recoil impulse.
⑨	Stock Weld:	Firm stock weld.
⑩	Shooter-Gun Angle:	Shooter body is nearly perpendicular with the gun-target line.

Figure 6-20. Reverse roll-over prone firing position

This page intentionally left blank.

Chapter 7

Aim

The functional element aim of the shot process is the continuous process of orienting the weapon correctly, aligning the sights, aligning on the target, and the application of the appropriate lead and elevation during a target engagement. Aiming is a continuous process conducted through pre-shot, shot, and post-shot, to effectively apply lethal fires in a responsible manner with accuracy and precision.

Aiming is the application of perfectly aligned sights on a specific part of a target. Sight alignment is the first and most important part of this process.

COMMON ENGAGEMENTS

7-1. The aiming process for engaging stationary targets consist of the following Soldier actions, regardless of the optic, sight, or magnification used by the aiming device:

- **Weapon orientation** – the direction of the weapon as it is held in a stabilized manner.
- **Sight alignment** – the physical alignment of the aiming device:
  - Iron sight/back-up iron sight and the front sight post.
  - Optic reticle.
  - Ballistic reticle (day or thermal).
- **Sight picture** – the target as viewed through the line of sight.
- **Point of aim (POA)** – the specific location where the line of sight intersects the target.
- **Desired point of impact (POI)** – the desired location of the strike of the round to achieve the desired outcome (incapacitation or lethal strike).

7-2. The aim of the weapon is typically applied to the largest, most lethal area of any target presented. Sights can be placed on target by using battlesight zero (BZ), *center of visible mass (CoVM)*. The center of visible mass is the initial point of aim on a target of what can be seen by the Soldier. It does not include what the target size is expected or anticipated to be. For example, a target located behind a car exposes its head. The center of visible mass is in the center of the head, not the estimated location of the center of the overall target behind the car.

Chapter 7

WEAPON ORIENTATION

7-3. The Soldier orients the weapon in the direction of the detected threat. Weapon orientation includes both the horizontal plane (azimuth) and the vertical plane (elevation). Weapon orientation is complete once the sight and threat are in the Soldier's field of view.

- **Horizontal weapons orientation** covers the frontal arc of the Soldier, spanning the area from the left shoulder, across the Soldier's front, to the area across the right shoulder (see figure 7-1).

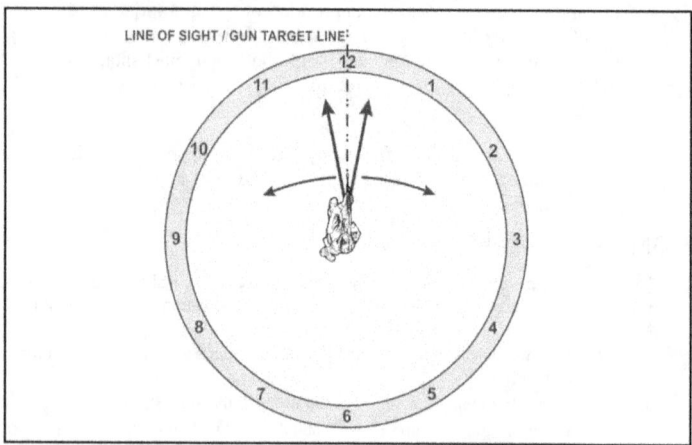

Figure 7-1. Horizontal weapon orientation example

Aim

- **Vertical weapons orientation** includes all the aspects of orienting the weapon at a potential or confirmed threat in elevation. This is most commonly applied in restricted, mountainous, or urban terrain where threats present themselves in elevated or depressed firing positions (see figure 7-2).

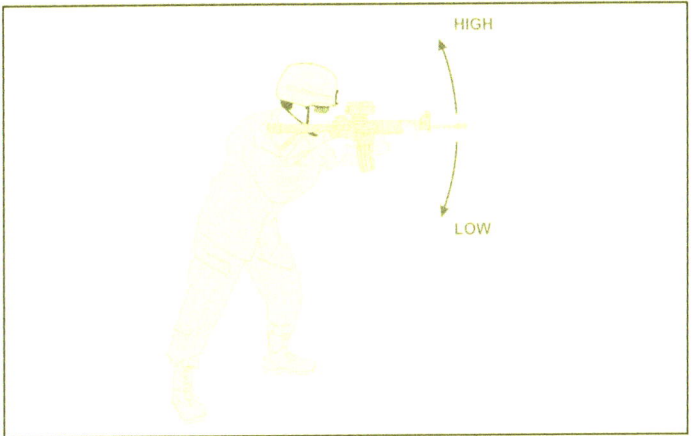

Figure 7-2. Vertical weapons orientation example

SIGHT ALIGNMENT

7-4. Sight alignment is the relationship between the aiming device and the firer's eye. The process used by a Soldier depends on the aiming device employed with the weapon.

- **Iron sight** – the relationship between the front sight post, rear sight aperture, and the firer's eye. The firer aligns the tip of the front sight post in the center of the rear aperture and his or her eye. The firer will maintain focus on the front sight post, simultaneously centering it in the rear aperture.
- **Optics** – the relationship between the reticle and the firer's eye and includes the appropriate eye relief, or distance of the Soldier's eye from the optic itself. Ensure the red dot is visible in the CCO, or a full centered field of view is achieved with no shadow on magnified optics
- **Thermal** – the relationship between the firer's eye, the eyepiece, and the reticle.
- **Pointers / Illuminators / Lasers** – the relationship between the firer's eye, the night vision device placement and focus, and the laser aiming point on the target.

Chapter 7

> *Note.* Small changes matter - 1/1000 of an inch deviation at the weapon can result in up to an 18 inch deviation at 300 meters.

7-5. The human eye can only focus clearly on one object at a time. To achieve proper and effective aim, the focus of the firer's eye needs to be on the front sight post or reticle (see figure 7-3). This provides the most accurate sight alignment for the shot process.

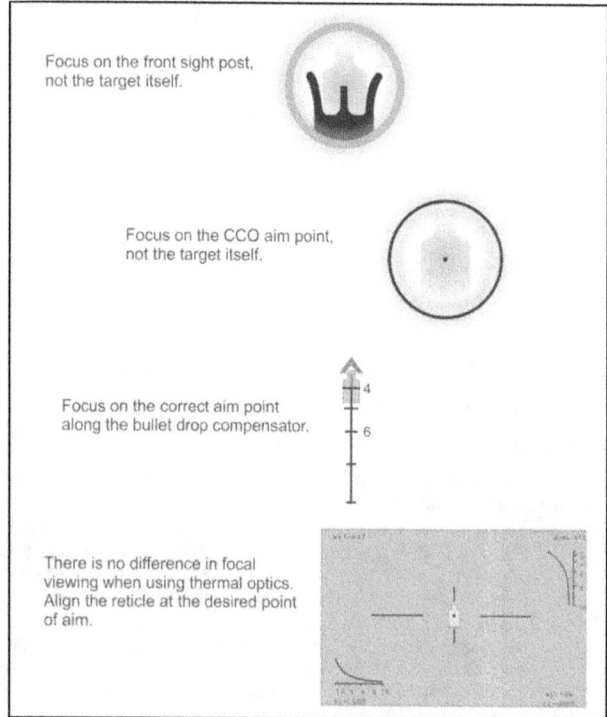

Figure 7-3. Front sight post/reticle aim focus

7-6. Firers achieve consistent sight alignment by resting the full weight of their head on the stock in a manner that allows their dominant eye to look through the center of the aiming or sighting device. If the firer's head placement is subjected to change during the firing process or between shots, the Soldier will experience difficultly achieving accurate shot groups.

Aim

SIGHT PICTURE

7-7. The sight picture is the placement of the aligned sights on the target itself. The Soldier must maintain sight alignment throughout the positioning of the sights. This is not the same as sight alignment.

7-8. There are two sight pictures used during the shot process; pre-shot and post-shot. Soldiers must remember the sight pictures of the shot to complete the overall shot process.
- Pre-shot sight picture – encompasses the original point of aim, sight picture, and any holds for target or environmental conditions.
- Post-shot sight picture – is what the Soldier must use as the point of reference for any sight adjustments for any subsequent shot.

POINT OF AIM

7-9. The point on the target that is the continuation of the line created by sight alignment. The point of aim is a point of reference used to calculate any hold the Soldier deems necessary to achieve the desired results of the round's impact.

7-10. For engagements against stationary targets, under 300 meters, with negligible wind, and a weapon that has a 200 meter or 300 meter confirmed zero, the point of aim should be the center of visible mass of the target. The point of aim does not include ANY hold-off or lead changes necessary.

DESIRED POINT OF IMPACT

7-11. The desired point of impact is the location where the Soldier wants the projectile to strike the target. Typically, this is the center of visible mass. At any range different from the weapon's zero distance, the Soldier's desired point of impact and their point of aim will not align. This requires the Soldier to determine the necessary hold-off to achieve the desired point of impact.

COMMON AIMING ERRORS

7-12. Orienting and aiming a weapon correctly is a practiced skill. Through drills and repetitions, Soldiers build the ability to repeat proper weapons orientation, sight alignment, and sight picture as a function of muscle memory.

7-13. The most common aiming errors include:
- **Non-dominant eye use** – The Soldier gets the greatest amount of visual input from their dominant eye. Eye dominance varies Soldier to Soldier. Some Soldier's dominant eye will be the opposite of the dominant hand. For example, a Soldier who writes with his right hand and learns to shoot rifles right handed might learn that his dominant eye is the left eye. This is called cross-dominant. Soldiers with strong cross-dominant eyes should consider firing using their dominant eye side while firing from their non-dominant hand side. Soldiers can be trained to fire from either side of the weapon, but may not be able to shoot effectively using their nondominant eye.

- **Incorrect zero** – regardless of how well a Soldier aims, if the zero is incorrect, the round will not travel to the desired point of impact without adjustment with subsequent rounds (see appendix C of this publication).
- **Light conditions** – limited visibility conditions contribute to errors aligning the sight, selecting the correct point of aim, or determining the appropriate hold. Soldiers may offset the effects of low light engagements with image intensifier (I2) optics, use of thermal optics, or the use of laser pointing devices with I2 optics.
- **Battlefield obscurants** – smoke, debris, and haze are common conditions on the battlefield that will disrupt the Soldier's ability to correctly align their sights, select the proper point of aim, or determine the correct hold for a specific target.
- **Incorrect sight alignment** – Soldiers may experience this error when failing to focus on the front sight post or reticle.
- **Incorrect sight picture** – occurs typically when the threat is in a concealed location, is moving, or sufficient winds between the shooter and target exist that are not accounted for during the hold determination process. This failure directly impacts the Soldier's ability to create and sustain the proper sight picture during the shot process.
- **Improper range determination** – will result in an improper hold at ranges greater than the zeroed range for the weapon.

COMPLEX ENGAGEMENTS

7-14. A complex engagement includes any shot that cannot use the *CoVM* as the point of aim to ensure a target hit. Complex engagements require a Soldier to apply various points of aim to successfully defeat the threat.

7-15. These engagements have an increased level of difficulty due to environmental, target, or shooter conditions that create a need for the firer to rapidly determine a ballistic solution and apply that solution to the point of aim. Increased engagement difficulty is typically characterized by one or more of the following conditions:
- **Target conditions**:
  - Range to target.
  - Moving targets.
  - Oblique targets.
  - Evasive targets.
  - Limited exposure targets.
- **Environmental conditions**:
  - Wind.
  - Angled firing.
  - Limited visibility.
- **Shooter conditions**:
  - Moving firing position.

Aim

- Canted weapon engagements.
- CBRN operations engagements.

7-16. Each of these firing conditions may require the Soldier to determine an appropriate aim point that is not the CoVM. This Soldier calculated aim point is called the **hold**. During any complex engagement, the Soldier serves as the ballistic computer during the shot process. The hold represents a refinement or alteration of the center of visible mass point of aim at the target to counteract certain conditions during a complex engagement for—

- Range to target.
- Lead for targets based on their direction and speed of movement.
- Counter-rotation lead required when the Soldier is moving in the opposite direction of the moving target.
- Wind speed, direction, and duration between the shooter and the target at ranges greater than 300 meters.
- Greatest lethal zone presented by the target to provide the most probable point of impact to achieve immediate incapacitation.

7-17. The Soldier will apply the appropriate aim (hold) based on the firing instances presented. Hold determinations will be discussed in two formats; immediate and deliberate.

7-18. All Soldiers must be familiar with the immediate hold determination methods. They should be naturally applied when the engagement conditions require. These determinations are provided in "target form" measurements, based on a standard E-type silhouette dimension, approximately 20 inches wide by 40 inches tall.

IMMEDIATE HOLD DETERMINATION

7-19. Immediate holds are based on the values of a "target form," where the increments shown *are sufficient* for rapid target hits without ballistic computations. The immediate hold determinations are not as accurate as the deliberate method, and are used for complex target engagements at less than 300 meters.

7-20. Immediate hold locations for azimuth (wind or lead): (See figure 7-4.)

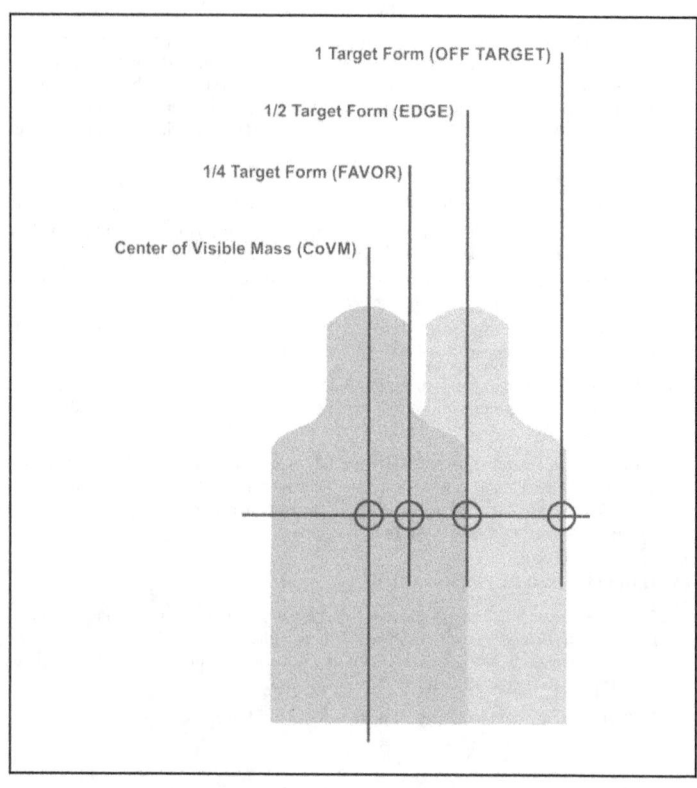

Figure 7-4. Immediate hold locations for windage and lead example

Aim

7-21. Immediate hold locations for elevation (range to target): (See figure 7-5.)

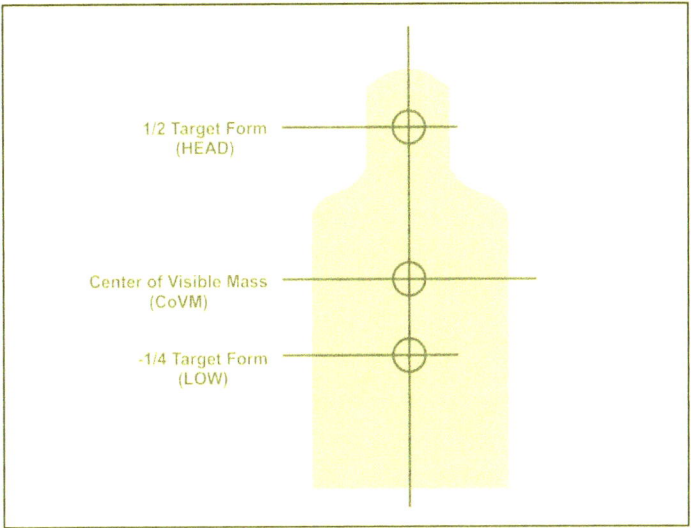

Figure 7-5. Immediate hold locations for elevation (range) example

DELIBERATE HOLD DETERMINATION

7-22. Deliberate hold points of aim are derived from applying the appropriate ballistic math computation. Deliberate hold determinations are required for precise shots beyond 300 meters for wind, extended range, moving, oblique, or evasive targets.

7-23. Deliberate holds for complex engagements are discussed in detail in appendix C, Complex Engagements. The deliberate math calculations are for advanced shooters within the formation and are not discussed within this chapter.

TARGET CONDITIONS

7-24. Soldiers must consider several aspects of the target to apply the proper point of aim on the target. The target's posture, or how it is presenting itself to the shooter, consists of—

- Range to target.
- Nature of the target.
- Nature of the terrain (surrounding the target).

RANGE TO TARGET

7-25. Rapidly determining an accurate range to target is critical to the success of the Soldier at mid and extended ranges. There are several range determination methods shooters should be confident in applying to determine the proper hold-off for pending engagements. There are two types of range determination methods, immediate and deliberate.

Immediate Range Determination

7-26. Immediate methods of range determination afford the shooter the most reliable means of determining the most accurate range to a given target. The immediate methods include—

- Close quarters engagements.
- Laser range finder.
- Front sight post method.
- Recognition method.
- 100-meter unit-of-measure method.

Close Quarters Engagements

7-27. Short-range engagements are probable in close terrain (such as urban or jungle) with engagement ranges typically less than 50 meters. Soldiers must be confident in their equipment, zero, and capabilities to defeat the threats encountered.

7-28. Employment skills include swift presentation and application of the shot process (such as quick acquisition of sight picture) to maintain overmatch. At close ranges, perfect sight alignment is not as critical to the accurate engagement of targets. The weapon is presented rapidly and the shot is fired with the front sight post placed roughly center mass on the desired target area. The front sight post must be in the rear sight aperture.

Note. If using iron sights when this type of engagement is anticipated, the large rear sight aperture (0-2) should be used to provide a wider field of view and detection of targets.

Laser Range Finder

7-29. Equipment like the AN/PSQ-23, STORM have an on-board laser range finder that is accurate to within +/- 5 meters. Soldiers with the STORM attached can rapidly

Aim

determine the most accurate range to target and apply the necessary hold-offs to ensure the highest probability of incapacitation, particularly at extended ranges.

Front Sight Post Method

7-30. The area of the target that is covered by the front sight post of the rifle can be used to estimate range to the target. By comparing the appearance of the rifle front sight post on a target at known distances, your shooters can establish a mental reference point for determining range at unknown distances. Because the apparent size of the target changes as the distance to the target changes, the amount of the target that is covered by the front sight post will vary depending upon its range. In addition, your shooter's eye relief and perception of the front sight post will also affect the amount of the target that is visible (see figure 7-6).

- Less Than 300 Meters. If the target is wider than the front sight post, you can assume that the target is less than 300 meters and can be engaged point of aim point of impact using your battle sight zero (BZO).
- Greater Than 300 Meters. The service rifle front sight post covers the width of a man's chest or body at approximately 300 meters. If the target is less than the width of the front sight post, you should assume the target is in excess of 300 meters. Therefore, your BZO cannot be used effectively.

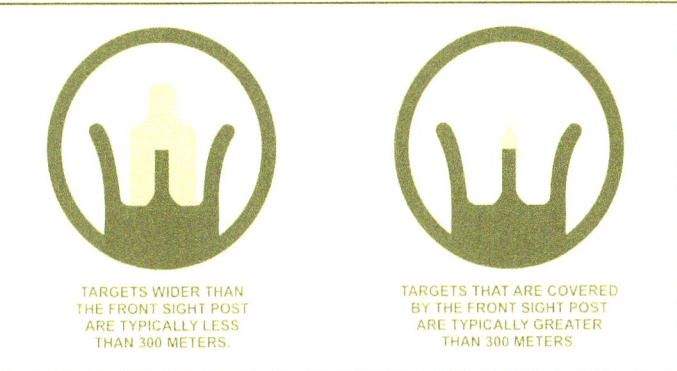

Figure 7-6. Front sight post method example

Recognition Method

7-31. When observing a target, the amount of detail seen at various ranges gives the shooter a solid indication of the range to target. Shooters should study and remember the appearance of a person when they are standing at 100 meters increments. During training, Soldiers should note the details of size and the characteristics of uniform and equipment for targets at those increments.

Chapter 7

7-32. Once Soldiers are familiar and memorize the characteristics of standing threats at 100 meter increments out to 500 meters, they should study the targets in a kneeling and then in the prone position. By comparing the appearance of these positions at known ranges from 100 meters to 500 meters, shooters can establish a series of mental images that will help determine range on unfamiliar terrain. They should also study the appearance of other familiar objects such as weapons and vehicles.

- **100 meters** – the target can be clearly observed in detail, and facial features can be distinguished.
- **200 meters** – the target can be clearly observed, although there is a loss of facial detail. The color of the skin and equipment is still identifiable.
- **300 meters** – the target has a clear body outline, face color usually remains accurate, but remaining details are blurred.
- **400 meters** – the body outline is clear, but remaining detail is blurred.
- **500 meters** – the body shape begins to taper at the ends. The head becomes indistinct from the shoulders.

100-meter Unit-of-Measure Method

7-33. To determine the total distance to the target using the 100 meter unit of measure method, shooters must visualize a distance of 100 meters (generally visualizing the length of a football field) on the ground. Soldiers then estimate how many of these units can fit between the shooter and the target.

7-34. The greatest limitation of the unit of measure method is that its accuracy is directly related to how much of the terrain is visible. This is particularly true at greater ranges. If a target appears at a range of 500 meters or more and only a portion of the ground between your shooter and the target can be seen, it becomes difficult to use the unit of measure method of range estimation with accuracy.

7-35. Proficiency in the unit of measure method requires constant practice. Throughout training, comparisons should be continually made between the range estimated by your shooter and the actual range as determined by pacing or other, more accurate measurement.

Immediate hold for Range to Target

7-36. Immediate range determination holds are based on the zero applied to the weapon. The 300 meter zero is the Army standard and works in all tactical situations, including close quarters combat. Figure 7-7, on page 7-13, shows the appropriate immediate holds for range to target based on the weapon's respective zero:

Aim

RANGE	HOLD	IRON SIGHT	CCO, M68
500 m	1 FORM OVER	USE BDC	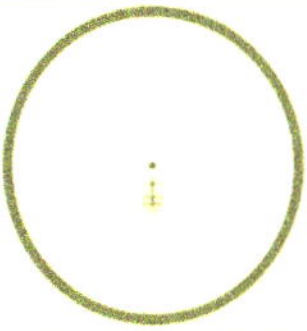
400 m	1/2 HEAD	USE BDC	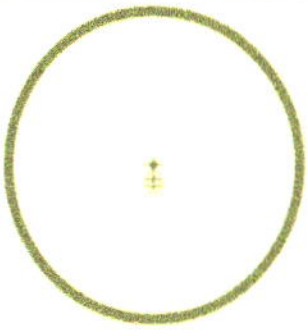
300 m	CoVM		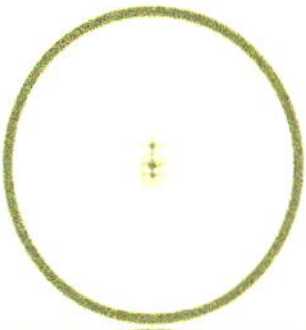
200 m	-1/4 LOW		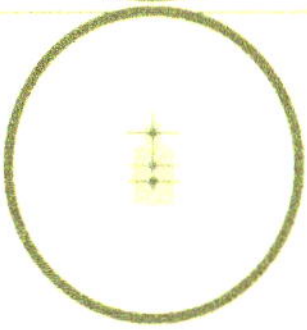
100 m	-1/4 LOW	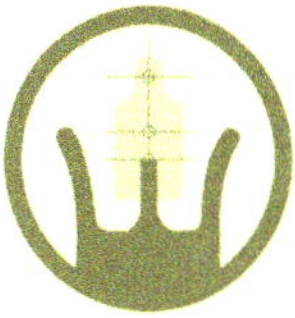	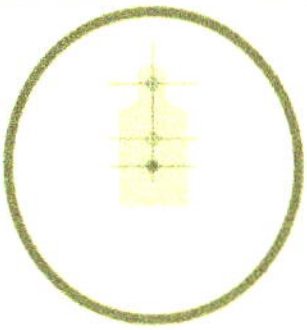

BDC - Bullet Drop Compensator

Figure 7-7. Immediate holds for range to target

Moving Targets

7-37. Moving targets are those threats that appear to have a consistent pace and direction. Targets on any battlefield will not remain stationary for long periods of time, particularly once a firefight begins. Soldiers must have the ability to deliver lethal fires at a variety of moving target types and be comfortable and confident in the engagement techniques. There are two methods for defeating moving targets; tracking and trapping.

Immediate hold for moving targets.

7-38. The immediate hold for moving targets includes an estimation of the speed of the moving target and an estimation of the range to that target. The immediate holds for all moving targets are shown below. (See figure 7-8.)

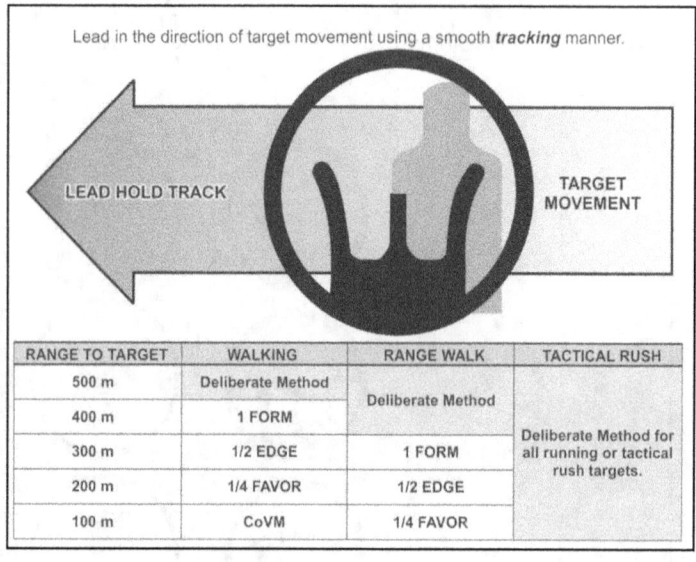

Figure 7-8. Immediate holds for moving targets example

Aim

OBLIQUE TARGETS

7-39. Threats that are moving diagonally toward or away from the shooter are called *oblique targets*. They offer a unique problem set to shooters where the target may be moving at a steady pace and direction; however, their oblique direction of travel makes them appear to move slower.

7-40. Soldiers should adjust their hold based on the angle of the target's movement from the gun-target line. The following guide will help Soldiers determine the appropriate change to the moving target hold to apply to engage the moving oblique threats (see figure 7-9).

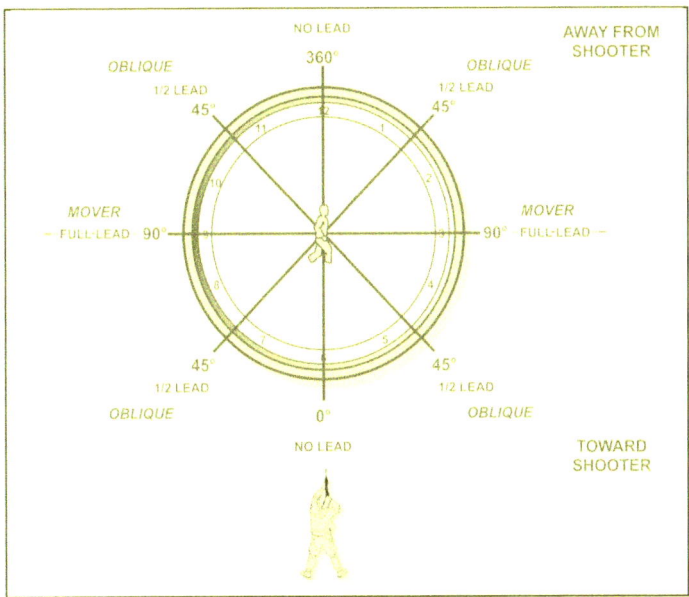

Figure 7-9. Oblique target example

ENVIRONMENTAL CONDITIONS

7-41. The environment can complicate the shooter's actions during the shot process with excessive wind or requiring angled firing limited visibility conditions. Soldiers must understand the methods to offset or compensate for these firing occasions, and be prepared to apply these skills to the shot process. This includes when multiple complex conditions compound the ballistic solution during the firing occasion.

WIND

7-42. Wind is the most common variable and has the greatest effect on ballistic trajectories, where it physically pushes the projectile during flight off the desired trajectory (see appendix B of this publication). The effects of wind can be compensated for by the shooter provided they understand how wind effects the projectile and the terminal point of impact. The elements of wind effects are—

- The time the projectile is exposed to the wind (range).
- The direction from which the wind is blowing.
- The velocity of the wind on the projectile during flight.

Wind Direction and Value

7-43. Winds from the left blow the projectile to the right, and winds from the right blow the projectile to the left. The amount of the effect depends on the time of (projectile's exposure) the wind speed and direction. To compensate for the wind, the firer must first determine the wind's direction and value.

7-44. The clock system can be used to determine the direction and value of the wind (See figure 7-10 on page 7-17). Picture a clock with the firer oriented downrange towards 12 o'clock.

7-45. Once the direction is determined, the value of the wind is next. The value of the wind is how much effect the wind will have on the projectile. Winds from certain directions have less effect on projectiles. The chart below shows that winds from 2 to 4°o'clock and 8 to 10 o'clock are considered full-value winds and will have the most effect on the projectile. Winds from 1, 5, 7, and 11 o'clock are considered half-value winds and will have roughly half the effect of a full-value wind. Winds from 6 and 12°o'clock are considered no-value winds and little or no effect on the projectile.

> **EXAMPLE**
>
> A 10-mph (miles per hour) wind blowing from the 1 o'clock direction would be a half-value wind and has the same effect as a 5 mph, full-value wind on the projectile.

Aim

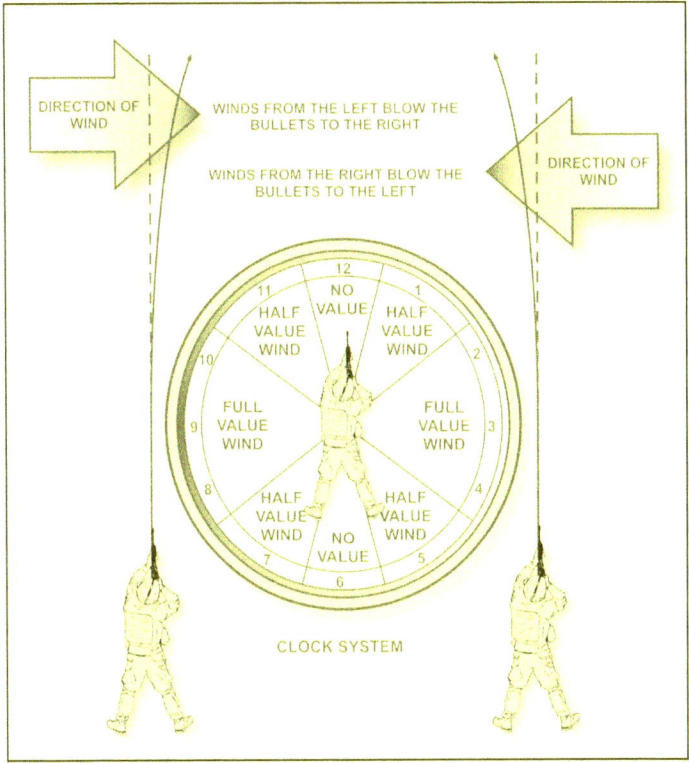

Figure 7-10. Wind value

7-46. The wind will push the projectile in the direction the wind is blowing (see figure 7-11). The amount of effects on the projectile will depend on the time of exposure, direction of the wind, and speed of the wind. To compensate for wind the Soldier uses a hold *in the direction of the wind (into the wind)*.

Chapter 7

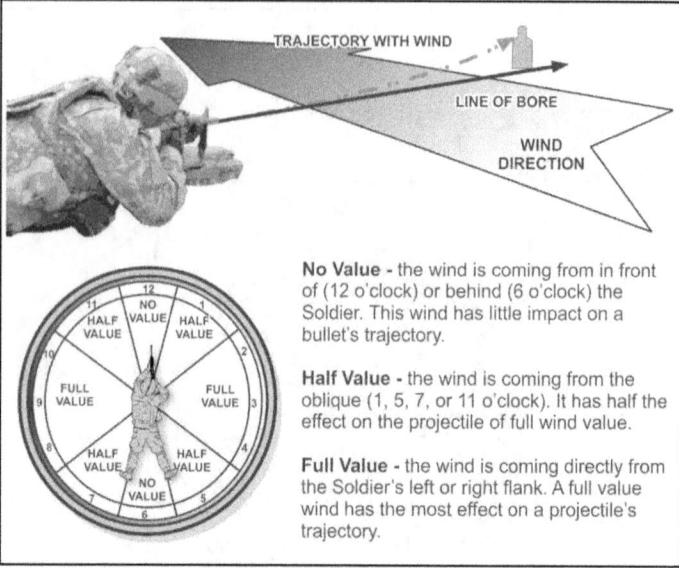

Figure 7-11. Wind effects

Wind Speed

7-47. Wind speeds can vary from the firing line to the target. Wind speed can be determined by taking an average of the winds blowing on the range. The firer's focus should be on the winds between the midrange point and the target. The wind at the one half to two thirds mark will have the most effect on the projectile since that is the point where most projectiles have lost a large portion of their velocity and are beginning to destabilize.

7-48. The Soldier can observe the movement of items in the environment downrange to determine the speed. Each environment will have different vegetation that reacts differently.

7-49. Downrange wind indicators include the following:
- 0 to 3 mph = Hardly felt, but smoke drifts.
- 3 to 5 mph = Felt lightly on the face.
- 5 to 8 mph = Keeps leaves in constant movement.

Aim

- 8 to 12 mph – Raises dust and loose paper.
- 12 to 15 mph – Causes small trees to sway.

7-50. The wind blowing at the Soldiers location may not be the same as the wind blowing on the way to the target.

Wind Estimation

7-51. Soldiers must be comfortable and confident in their ability to judge the effects of the wind to consistently make accurate and precise shots. Soldiers will use wind indicators between the Soldier and the target that provide windage information to develop the proper compensation or hold-off.

7-52. To estimate the effects of the wind on the shot, Soldiers need to determine three windage factors:

- Velocity (speed).
- Direction.
- Value.

Immediate Wind Hold

7-53. Using a hold involves changing the point of aim to compensate for the wind drift. For example, if wind causes the bullet to drift 1/2 form to the left, the aiming point must be moved 1/2 form to the right. (See figure 7-12.)

Chapter 7

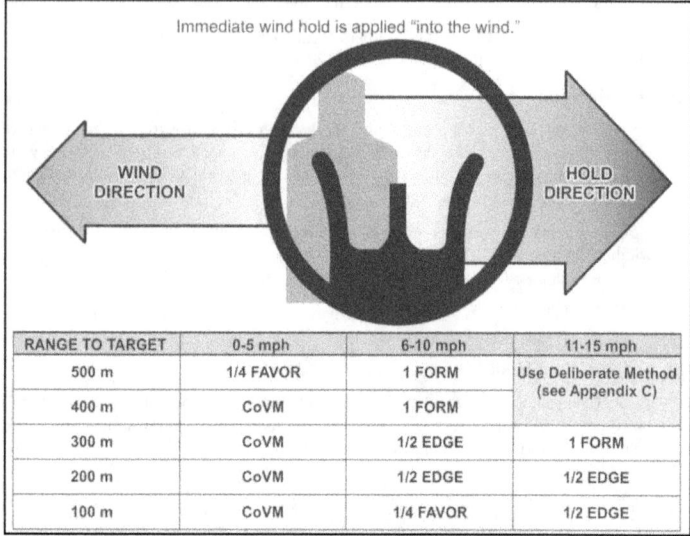

Figure 7-12. Wind hold example

7-54. Firers must adjust their points of aim into the wind to compensate for its effects. If they miss a distant target and wind is blowing from the right, they should aim to the right for the next shot. A guide for the initial adjustment is to split the front sight post on the edge of the target facing the wind.

7-55. Newly assigned Soldiers should aim at the target's center of visible mass for the first shot, and then adjust for wind when they are confident that wind caused the miss. Experienced firers should apply the appropriate hold for the first shot, but should follow the basic rule—when in doubt, aim at the center of mass.

LIMITED VISIBILITY

7-56. Soldiers must be lethal at night and in limited visibility conditions, as well as during the day. That lethality depends largely on whether Soldier can fire effectively with today's technology: night vision devices (NVDs), IR aiming devices, and TWSs.

7-57. Limited visibility conditions may limit the viewable size of a threat, or cause targets to be lost after acquisition. In these situations, Soldiers may choose to apply a hold for where a target is *expected* to be rather than wait for the target to present itself for a more refined reticle lay or sight picture.

7-58. Soldiers may switch between optics, thermals, and pointers to refine their point of aim. To rapidly switch between aiming devices during operations in limited visibility, the Soldier must ensure accurate alignment, boresighting, and zeroing of all associated equipment. Confidence in the equipment is achieved through drills related to changing the aiming device during engagements, executing repetitions with multiple pieces of equipment, and practicing nonstandard engagement techniques using multiple aiming devices in tandem (IR pointer with NVDs, for example).

SHOOTER CONDITIONS

7-59. The ability to aim properly while the shooter is moving, has the weapon canted (tilted to one side or the other), or is fighting in a CBRN environment creates additional difficulties to achieve the appropriate point of aim. These shooter conditions can be mitigated to ensure effective point of aim and target defeat.

TACTICAL MOVEMENT

7-60. A Soldier moving tactically in any direction and attempting to engage a target may require an increase or decrease in the lead applied to a target. The following rules apply:

- A Soldier is moving in the same direction as the target, or the target is stationary, the Soldier must apply counter-lead to offset his forward movement. The counter-lead (or counter-rotation) is based upon the range to target, the speed of the Soldier, and the speed of the target. Typically, this movement negates the need for any lead hold-off.
- A Soldier moving in an opposite direction of the threat, the Soldier applies twice the amount of lead.

CANTED WEAPON

7-61. If the weapon must be tilted (canted) in one direction or another to engage a target (like in the case of the prone, roll-over position), the strike of the bullet will be in the direction of the canted weapon and low. When firing a canted weapon, the elevation becomes the azimuth, and the azimuth becomes the elevation in relation to the aim point.

Rule of Thumb for a Canted Weapon

Aim to the magazine side and slightly high over the bore based on the angle of the cant.

Close Range

7-62. At close range, the effects of cant are specific to the line of sight and the axis of the bore. Soldiers should apply the offset to the target based on the angle of the cant.

Extended Range

7-63. The general rule is to apply the aim point in an equal amount in the opposite direction of the cant to ensure the highest probability of hit.

Aim

COMPOUND CONDITIONS

7-64. When combining difficult target firing occasion information, Soldiers can apply the rules specific to the situation together to determine the appropriate amount of hold-off to apply.

7-65. The example below shows the application of different moving target directions with varying speed directions (see figure 7-13). This is a general example to provide the concept of applying multiple hold-off information to determine complex ballistic solutions for an engagement. This same concept is applied to immediate and deliberate methods of determining hold.

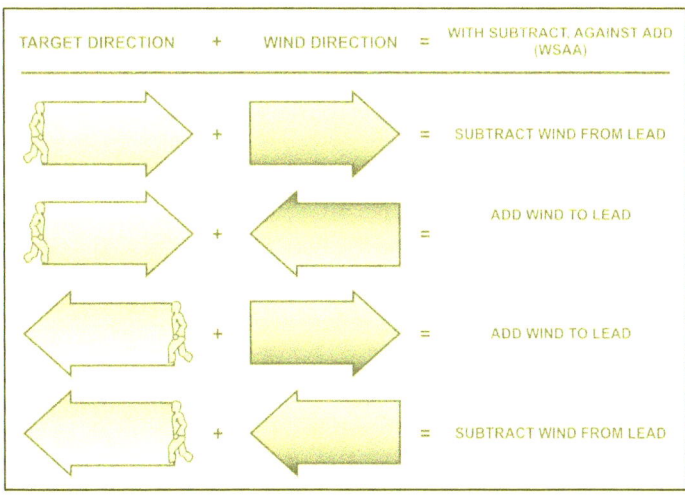

Figure 7-13. Compound wind and lead determination example

This page intentionally left blank.

Chapter 8
Control

The control element of employment considers all the conscious actions of the Soldier before, during, and after the shot process that the Soldier's specifically in control of. It incorporates the Soldier as a function of safety, as well as the ultimate responsibility of firing the weapon.

Proper trigger control, without disturbing the sights, is the most important aspect of control and the most difficult to master.

Combat is the ultimate test of a Soldier's ability to apply the functional elements of the shot process and firing skills. Soldiers must apply the employment skills mastered during training to all combat situations (for example, attack, assault, ambush, or urban operations). Although these tactical situations present problems, the application of the functional elements of the shot process require two additions: changes to the rate of fire and alterations in weapon/target alignment. This chapter discusses the engagement techniques Soldiers must adapt to the continuously changing combat engagements.

8-1. When firing individual weapons, the Soldier is the weapon's fire control system, ballistic computer, stabilization system, and means of mobility. Control refers to the Soldier's ability to regulate these functions and maintain the discipline to execute the shot process at the appropriate time.

8-2. Regardless of how well trained or physically strong a Soldier is, a wobble area (or arc of movement) is present, even when sufficient physical support of the weapon is provided. The arc of movement (AM) may be observed as the sights moving in a W shape, vertical (up and down) pulses, circular, or horizontal arcs depending on the individual Soldier, regardless of their proficiency in applying the functional elements. The wobble area or arc of movement is the extent of lateral horizontal and front-to-back variance in the movement that occurs in the sight picture (see figure 8-1).

Chapter 8

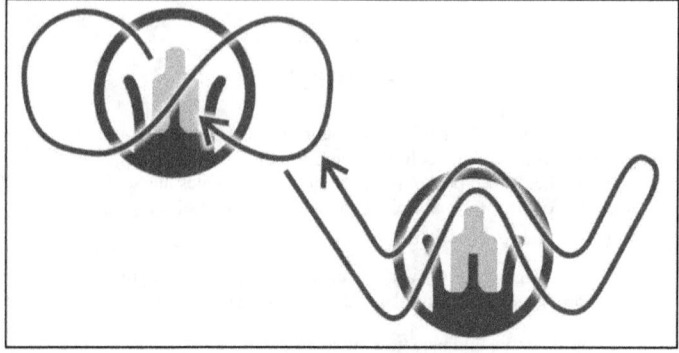

Figure 8-1. Arc of movement example

8-3. The control element consists of several supporting Soldier functions, and include all the actions to minimize the Soldier's induced arc of movement. Executed correctly, it provides for the best engagement window of opportunity to the firer. The Soldier physically maintains positive control of the shot process by managing—

- Trigger control.
- Breathing control.
- Workspace.
- Calling the shot (firing or shot execution).
- Follow-through.

TRIGGER CONTROL

8-4. Trigger control is the act of firing the weapon while maintaining proper aim and adequate stabilization until the bullet leaves the muzzle. Trigger control and the shooter's position work together to allow the sights to stay on the target long enough for the shooter to fire the weapon and bullet to exit the barrel.

8-5. Stability and trigger control complement each other and are integrated during the shot process. A stable position assists in aiming and reduces unwanted movements during trigger squeeze without inducing unnecessary movement or disturbing the sight picture. A smooth, consistent trigger squeeze, regardless of speed, allows the shot to fire at the Soldier's moment of choosing. When both a solid position and a good trigger squeeze are achieved, any induced shooting errors can be attributed to the aiming process for refinement.

8-6. Smooth trigger control is facilitated by placing the finger where it naturally lays on the trigger. Natural placement of the finger on the trigger will allow for the best mechanical advantage when applying rearward pressure to the trigger.

Control

- **Trigger finger placement** – the trigger finger will lay naturally across the trigger after achieving proper grip (see figure 8-2). There is no specified point on the trigger finger that must be used. It will not be the same for all Soldiers due to different size hands. This allows the Soldier to engage the trigger in the most effective manner
- **Trigger squeeze** – The Soldier pulls the trigger in a smooth consistent manner adding pressure until the weapon fires. Regardless of the speed at which the Soldier is firing the trigger control will always be smooth.
- **Trigger reset** – It is important the Soldier retains focus on the sights while resetting the trigger.

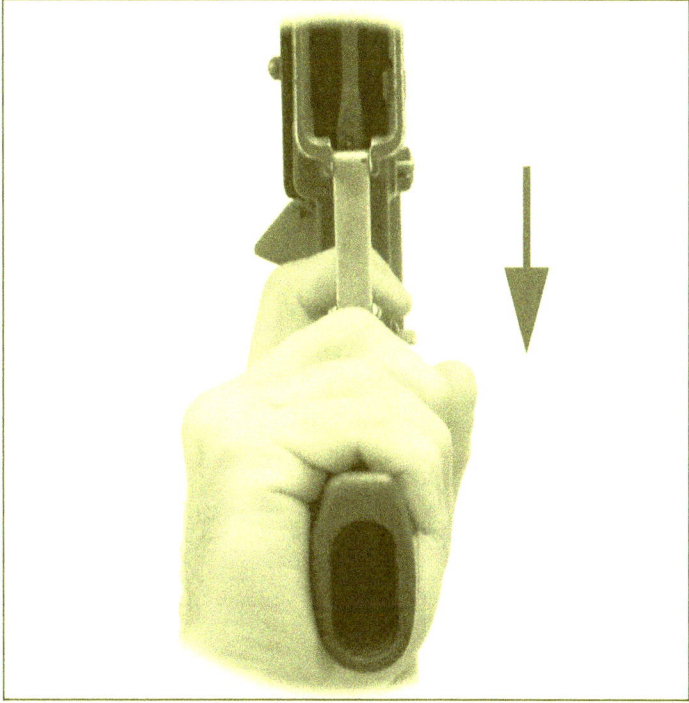

Figure 8-2. Natural trigger finger placement

BREATHING CONTROL

8-7. During the shot process, the shooter controls their breathing to reduce the amount of movement of the weapon. During training, the Soldier will learn a method of breathing control that best suits their shooting style and preference. Breathing control is the relationship of the respiratory process (free or under stress) and the decision to execute the shot with trigger squeeze.

8-8. Breathing induces unavoidable body movement that contribute to wobble or the arc of movement (AM) during the shot process. Soldiers cannot completely eliminate all motion during the shot process, but they can significantly reduce its effects through practice and technique. Firing on the natural pause is a common technique used during grouping and zeroing.

8-9. Vertical dispersion during grouping is most likely not caused by breathing but by failure to maintain proper aiming and trigger control. Refer to appendix E of this publication for proper target analysis techniques.

WORKSPACE MANAGEMENT

8-10. The workspace is a spherical area, 12 to 18 inches in diameter centered on the Soldier's chin and approximately 12 inches in front of their chin. The workspace is where the majority of weapons manipulations take place. (See figure 8-3 on page 8-5.)

8-11. Conducting manipulations in the workspace allows the Soldier to keep his eyes oriented towards a threat or his individual sector of fire while conducting critical weapons tasks that require hand and eye coordination. Use of the workspace creates efficiency of motion by minimizing the distance the weapon has to move between the firing position to the workspace and return to the firing position.

8-12. Location of the workspace will change slightly in different firing positions. There are various techniques to use the workspace. Some examples are leaving the butt stock in the shoulder, tucking the butt stock under the armpit for added control of the weapon, or placing the butt stock in the crook of the elbow.

8-13. Workspace management includes the Soldier's ability to perform the following functions:

- **Selector lever** – to change the weapon's status from safe to semiautomatic, to burst/automatic from any position.

Note. Some models will have ambidextrous selectors.

- **Charging handle** – to smoothly use the charging handle during operation. This includes any corrective actions to overcome malfunctions, loading, unloading, or clearing procedures.
- **Bolt catch** – to operate the bolt catch mechanism on the weapon during operations.
- **Ejection port** – closing the ejection port cover to protect the bolt carrier assembly, ammunition, and chamber from external debris upon completion

Control

of an engagement. This includes observation of the ejection port area during malfunctions and clearing procedures.
- **Magazine catch** – the smooth functioning of the magazine catch during reloading procedures, clearing procedures, or malfunction corrective actions.
- **Chamber check** – the sequence used to verify the status of the weapon's chamber.
- **Forward assist** – the routine use of the forward assist assembly of the weapon during loading procedures or when correcting malfunctions.

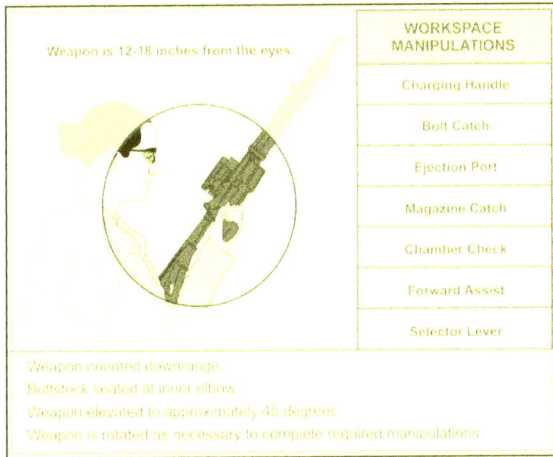

Figure 8-3. Workspace example

CALLING THE SHOT

8-14. Knowing precisely where the sights are when the weapon discharges is critical for shot analysis. Errors such as flinching or jerking of the trigger can be seen in the sights before discharge.

8-15. Calling a shot refers to a firer stating exactly where he thinks a single shot strikes by recalling the sights relationship to the target when the weapon fired. This is normally expressed in clock direction and inches from the desired point of aim.

8-16. The shooter is responsible for the point of impact of every round fired from their weapon. This requires the Soldier to ensure the target area is clear of friendly and neutral actors, in front of and behind the target. Soldiers must also be aware of the environment the target is positioned in, particularly in urban settings—friendly or neutral actors may be present in other areas of a structure that the projectile can pass through.

RATE OF FIRE

8-17. The shooter must determine *how* to engage the threat with the weapon, on the current shot as well as subsequent shots. Following the direction of the team leader, the Soldier controls the rate of fire to deliver consistent, lethal, and precise fires against the threat.

SLOW SEMIAUTOMATIC FIRE

8-18. Slow semiautomatic fire is moderately paced at the discretion of the Soldier, typically used in a training environment or a secure defensive position at approximately 12 to 15 rounds per minute. All Soldiers learn the techniques of slow semiautomatic fire during their introduction to the service rifle during initial entry training. This type of firing provides the Soldier the most time to focus on the functional elements in the shot process and reinforces all previous training.

RAPID SEMIAUTOMATIC FIRE

8-19. Rapid semiautomatic fire is approximately 45 rounds per minute and is typically used for multiple targets or combat scenarios where the Soldier does not have overmatch of the threat. Soldiers should be well-trained in all aspects of slow semiautomatic firing before attempting any rapid semiautomatic fire training.

8-20. Those who display a lack of knowledge of employment skills should not advance to rapid semiautomatic fire training until these skills are learned and mastered.

AUTOMATIC OR BURST FIRE

8-21. Automatic or burst fire is when the Soldier is required to provide suppressive fires with accuracy, and the need for precise fires, although desired, is not as important. Automatic or burst fires drastically decrease the probability of hit due to the rapid succession of recoil impulses and the inability of the Soldier to maintain proper sight alignment and sight picture on the target.

8-22. Soldiers should be well-trained in all aspects of slow semiautomatic firing before attempting any automatic training.

Control

FOLLOW-THROUGH

8-23. Follow-through is the continued mental and physical application of the functional elements of the shot process after the shot has been fired. The firer's head stays in contact with the stock, the firing eye remains open, the trigger finger holds the trigger back through recoil and then lets off enough to reset the trigger, and the body position and breathing remain steady.

8-24. Follow-through consists of all actions controlled by the shooter after the bullet leaves the muzzle. It is required to complete the shot process. These actions are executed in a general sequence:

- **Recoil management.** This includes the bolt carrier group recoiling completely and returning to battery.
- **Recoil recovery.** Returning to the same pre-shot position and reacquiring the sight picture. The shooter should have a good sight picture before and after the shot.
- **Trigger/Sear reset.** Once the ejection phase of the cycle of function is complete, the weapon initiates and completes the cocking phase. As part of the cocking phase, all mechanical components associated with the trigger, disconnect, and sear are reset. Any failures in the cocking phase indicate a weapon malfunction and require the shooter to take the appropriate action. The shooter maintains trigger finger placement and releases pressure on the trigger until the sear is reset, demonstrated by a metallic click. At this point the sear is reset and the trigger pre-staged for a subsequent or supplemental engagement if needed.
- **Sight picture adjustment.** Counteracting the physical changes in the sight picture caused by recoil impulses and returning the sight picture onto the target aiming point.
- **Engagement assessment.** Once the sight picture returns to the original point of aim, the firer confirms the strike of the round, assesses the target's state, and immediately selects one of the following courses of action:
  - **Subsequent engagement.** The target requires additional (subsequent) rounds to achieve the desired target effect. The shooter starts the pre-shot process.
  - **Supplemental engagement.** The shooter determines the desired target effect is achieved and another target may require servicing. The shooter starts the pre-shot process.
  - **Sector check.** All threats have been adequately serviced to the desired effect. The shooter then checks his sector of responsibility for additional threats as the tactical situation dictates. The unit's SOP will dictate any vocal announcements required during the post-shot sequence.
  - **Correct Malfunction.** If the firer determines during the follow-through that the weapon failed during one of the phases of the cycle of function, they make the appropriate announcement to their team and immediately execute corrective action.

Chapter 8

MALFUNCTIONS

8-25. When any weapon fails to complete any phase of the cycle of function correctly, a malfunction has occurred. When a malfunction occurs, the Soldier's priority remains to defeat the target as quickly as possible. The malfunction, Soldier capability, and secondary weapon capability determine if, when, and how to transition to a secondary weapon system.

8-26. The Soldier controls which actions must be taken to ensure the target is defeated as quickly as possible based on secondary weapon availability and capability, and the level of threat presented by the range to target and its capability:

- **Secondary weapon can defeat the threat.** Soldier transitions to secondary weapon for the engagement. If no secondary weapon is available, announce their status to the small team, and move to a covered position to correct the malfunction.
- **Secondary weapon cannot defeat the threat**. Soldiers quickly move to a covered position, announce their status to the small team, and execute corrective action.
- **No secondary weapon.** Soldiers quickly move to a covered position, announce their status to the small team, and execute corrective action.

8-27. The end state of any of corrective action is a properly functioning weapon. Typically, the phase where the malfunction occurred within the cycle of function identifies the general problem that must be corrected. From a practical, combat perspective, malfunctions are recognized by their symptoms. Although some symptoms do not specifically identify a single point of failure, they provide the best indication on which corrective action to apply.

8-28. To overcome the malfunction, the Soldier must first avoid over analyzing the issue. The Soldier must train to execute corrective actions immediately without hesitation or investigation during combat conditions.

8-29. There are two general types of corrective action:

- **Immediate action** – simple, rapid actions or motions taken by the Soldier to correct basic disruptions in the cycle of function of the weapon. Immediate action is taken when a malfunction occurs such that the trigger is squeeze and the hammer falls with an audible "click."
- **Remedial action** – a skilled, technique that must be applied to a specific problem or issue with the weapon that will not be corrected by taking immediate action. Remedial action is taken when the cycle of function is interrupted where the trigger is squeezed and either has little resistance during the squeeze ("mush") or the trigger cannot be squeezed.

8-30. No single corrective action solution will resolve *all* or *every* malfunction. Soldiers need to understand what failed to occur, as well as any specific sounds or actions of the weapon in order to apply the appropriate correction measures.

Control

8-31. Immediate action can correct rudimentary failures during the cycle of function:
- **Failure to fire** – is when a round is locked into the chamber, the weapon is ready to fire, the select switch is placed on SEMI or BURST / AUTO, and the trigger is squeezed, the hammer falls (audible click), and the weapon does not fire.
- **Failure to feed** – is when the bolt carrier assembly is expected to move return back into battery but *is prevented from moving all the way forward*. A clear gap can be seen between the bolt carrier assembly and the forward edge of the ejection port. This failure may cause a stove pipe or a double feed (see below).
- **Failure to chamber** – when the round is being fed into the chamber, but the bolt carrier assembly does not fully seat forward, failing to chamber the round and lock the bolt locking lugs with the barrel extension's corresponding lugs.
- **Failure to extract** – when either automatically or manually, the extractor loses its grip on the cartridge case or the bolt seizes movement rearward during extraction that leaves the cartridge case partially removed or fully seated.
- **Failure to eject** – occurs when, either automatically or manually, a cartridge case is extracted from the chamber fully, but does not leave the upper receiver through the ejection port.

8-32. Remedial action requires the Soldier to quickly identify one of four issues and apply a specific technique to correct the malfunction. Remedial action is required to correct the following types of malfunctions or symptoms:
- **Immediate action fails to correct symptom** – when a malfunction occurred that initiated the Soldier to execute immediate action and multiple attempts failed to correct the malfunction. A minimum of two cycles of immediate action should have been completed; first, without a magazine change, and the second with a magazine change.
- **Stove pipe** – can occur when either a feeding cartridge or an expended cartridge case is pushed sideways during the cycle of function causing that casing to stop the forward movement of the bolt carrier assembly and lodge itself between the face of the bolt and the ejection port.
- **Double feed** – occurs when a round is chambered and not fired and a subsequent round is being fed without the chamber being clear.
- **Bolt override** – is when the bolt fails to push a new cartridge out of the magazine during feeding or chambering, causing the bolt to ride on top of the cartridge.
- **Charging handle impingement** – when a round becomes stuck between the bolt assembly and the charging handle where the charging handle is not in the forward, locked position.

8-33. Although there are other types of malfunctions or disruptions to the cycle of function, those listed above are the most common. Any other malfunction will require additional time to determine the true point of failure and an appropriate remedy.

> *Note.* When malfunctions occur in combat, the Soldier must announce STOPPAGE or another similar term to their small unit, quickly move to a covered location, and correct the malfunction as rapidly as possible. If the threat is too close to the Soldier or friendly forces, and the Soldier has a secondary weapon, the Soldier should immediately transition to secondary to defeat the target prior to correcting the malfunction.

RULES FOR CORRECTING A MALFUNCTION

8-34. To clear a malfunction, the Soldier must—

- **Apply Rule #1.** Soldiers must remain coherent of their weapon and continue to treat their weapon as if it is loaded when correcting malfunctions.
- **Apply Rule #2.** Soldiers must ensure the weapon's orientation is appropriate for the tactical situation and not flag other friendly forces when correcting malfunctions.
- **Apply Rule #3.** Take the trigger finger off the trigger, keep it straight along the lower receiver placed outside of the trigger guard.
- **Do not attempt to place the weapon on SAFE** (unless otherwise noted). Most stoppages will not allow the weapon to be placed on safe because the sear has been released or the weapon is out of battery. Attempting to place the weapon on SAFE will waste time and potentially damage the weapon.
- **Treat the symptom.** Each problem will have its own specific symptoms. By reacting to what the weapon is "telling" the Soldier, they will be able to quickly correct the malfunction.
- **Maintain focus on the threat.** The Soldier must keep their head and eyes looking downrange at the threat, not at the weapon. If the initial corrective action fails to correct the malfunction, the Soldier must be able to quickly move to the next most probable corrective action.
- **Look last.** Do not look and analyze the weapon to determine the cause of the malfunction. Execute the drill that has the highest probability of correcting the malfunction.
- **Check the weapon.** Once the malfunction is clear and the threat is eliminated, deliberately check the weapon when in a covered location for any potential issues or contributing factors that caused the malfunction and correct them.

Control

Perform Immediate Action

8-35. To perform immediate action, the Soldier instinctively:
- Hears the hammer fall with an audible "click."
- Taps the bottom of the magazine firmly.
- Rapidly pulls the charging handle and releases to extract / eject the previous cartridge and feed, chamber, and lock a new round.
- Reassess by continuing the shot process.

Note. If a malfunction continues to occur with the same symptoms, the Soldier will remove the magazine and insert a new loaded magazine, then repeat the steps above.

Perform Remedial Action

8-36. To perform remedial action, the Soldier must have a clear understanding of where the weapon failed during the cycle of function. Remedial action executed when one of the following conditions exist:
- Immediate action does not work after two attempts.
- The trigger refuses to be squeezed.
- The trigger feels like "mush" when squeezed.

8-37. When one of these three symptoms exist, the Soldier looks into the chamber area through the ejection port to quickly assess the type of malfunction. Once identified, the Soldier executes actions to "reduce" the symptom by removing the magazine and attempting to clear the weapon. Once complete, visually inspect the chamber area, bolt face, and charging handle. Then, complete the actions for the identified symptom:
- **Stove pipe** – the Soldier must remove the magazine, clear the weapon, confirm the chamber area is clear, secure a new loaded magazine into the magazine well, and chamber and lock a round.
- **Double-feed** - the Soldier must remove the magazine, clear the weapon, confirm the chamber area is clear, secure a new loaded magazine into the magazine well, and chamber and lock a round.
- **Bolt override** – the Soldier attempts to clear the weapon. If unsuccessful, apply the mortar technique to clear the round out of the weapon.
- **Charging handle impingement** – the Soldier attempts to clear the weapon. If unsuccessful, apply the mortar technique to clear the round out of the weapon.

Note. The "mortar" technique requires instruction to perform correctly without damaging the weapon. This technique is specifically not listed in this manual due to the potential to damage equipment and personnel. If required during combat, Soldier must announce the malfunction to the team and seek cover as the tactical situation permits.

Chapter 8

CORRECTING MALFUNCTIONS

8-38. Figure 8-4 below provides a simple mental flow chart to rapidly overcome malfunctions experienced during the shot process.

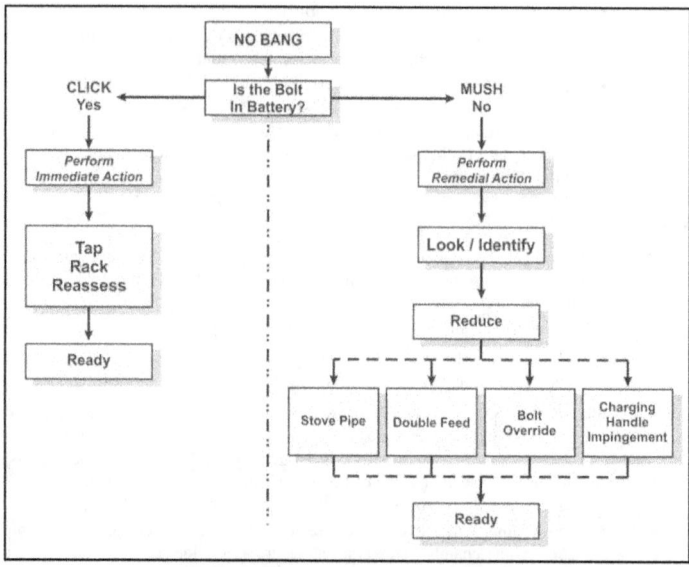

Figure 8-4. Malfunction corrective action flow chart

Control

COOK-OFF

8-39. Rapid and continuous firing of several magazines in sequence without cooling, will severely elevate chamber temperatures. While unlikely this elevated temperature may cause a malfunction known as a "cook-off". A "cook-off" may occur while the round is locked in the chamber, due to excessive heating of the ammunition. Or the rapid exposure to the cooler air outside of the chamber, due in part to the change in pressure.

8-40. If the Soldier determines that he has a potential "cook-off" situation he should leave the weapon directed at the target, or in a known safe direction, and follow proper weapons handling procedures, until the barrel of the weapon has had time to cool. If the chambered round has not been locked in the chamber for 10 seconds, it should be ejected as quickly as possible. If the length of time is questionable or known to be longer than 10 seconds and it is tactically sound, the Soldier should follow the above procedures until the weapon is cooled. If it is necessary to remove the round before the weapon has time to cool, the Soldier should do so with care as the ejected round may detonate due to rapid cooling in open air.

WARNING

Ammunition "cook-off" is not likely in well maintained weapons used within normal training and combat parameters.

Soldiers and unit leadership need to consider the dangers of keeping rounds chambered in weapons that have elevated temperatures due to excessive firing. Or clearing ammunition that has the potential to cook-off when exposed to the cooler air outside of the chamber.

Exposure to the colder air outside of the chamber has the potential to cause the "cook-off" of ammunition. Keeping ammunition chambered in severely elevated temperatures also has the potential to cause the "cook-off" of ammunition.

Note. For more information about troubleshooting malfunctions and replacing components, see organizational and direct support maintenance publications and manuals.

TRANSITION TO SECONDARY WEAPON

8-41. A secondary weapon, such as a pistol, is the most efficient way to engage a target at close quarters when the primary weapon has malfunctioned. The Soldier controls which actions must be taken to ensure the target is defeated as quickly as possible based on the threat presented.

8-42. The firer transitions by taking the secondary weapon from the HANG or HOLSTERED position to the READY UP position, reacquiring the target, and resuming the shot process as appropriate.

8-43. Refer to the appropriate secondary weapon's training publications for the specific procedures to complete the transition process.

Chapter 9

Movement

The movement functional element is the process of the Soldier moving tactically during the engagement process. It includes the Soldier's ability to move laterally, forward, diagonally, and in a retrograde manner while maintaining stabilization, appropriate aim, and control of the weapon.

Proper application of the shot process during movement is vital to combat operations. The most complex engagements involve movement of both Soldier and the adversary. The importance of sight alignment and trigger control are at their highest during movement. The movement of the Solider degrades stability, the ability to aim, and creates challenges to proper trigger control.

MOVEMENT TECHNIQUES

9-1. Tactical movement of the Soldier is classified in two ways: vertical and horizontal. Each require specific considerations to maintain and adequately apply the other functional elements during the shot process.

9-2. **Vertical movements** are those actions taken to change their firing posture or negotiate terrain or obstacles while actively seeking, orienting on, or engaging threats. Vertical movements include actions taken to—
- Change between any of the primary firing positions; standing, crouched, kneeling, sitting, or prone.
- Negotiate stairwells in urban environments.
- Travel across inclined or descending surfaces, obstacles, or terrain.

9-3. **Horizontal movements** are actions taken to negotiate the battlefield while actively seeking, orienting on, or engaging threats. There are eight horizontal movement techniques while maintaining weapon orientation on the threat—
- **Forward** – movement in a direction directly toward the adversary.
- **Retrograde** – movement rearward, in a direction away from the threat while maintaining weapon orientation on the threat.
- **Lateral right/left** – lateral, diagonal, forward, or retrograde movement to the right or left.
- **Turning left/right/about** – actions taken by the Soldier to change the weapon orientation left/right or to the rear, followed by the Soldier's direction of travel turning to the same orientation.

Chapter 9

FORWARD MOVEMENT

9-4. Forward movement is continued progress in a direction toward the adversary or route of march. This is the most basic form of movement during an engagement.

TECHNIQUE

9-5. During forward movement,—
- Roll the foot heel to toe to best provide a stable firing platform.
- Shooting while moving should be very close to the natural walking gait and come directly from the position obtained while stationary.
- Keep the weapon at the ready position. Always maintain awareness of the surroundings, both to your left and right, at all times during movement.
- Maintain an aggressive position.
- The feet should almost fall in line during movement. This straight-line movement will reduce the arc of movement and visible "bouncing" of the sight picture.
- Keep the muzzle of the weapon facing down range toward the expected or detected threat.
- Keep the hips as stationary as possible. Use the upper body as a turret, twisting at the waist, maintaining proper platform with the upper body.

RETROGRADE MOVEMENT

9-6. Retrograde movement is where the orientation of the weapon remains to the Soldier's front while the Soldier methodically moves rearward.

TECHNIQUE

9-7. During retrograde movement, the Soldier should—
- Take only one or two steps that will open the distance or reposition the feet.
- Place the feet in a toe to heel manner and drop the center body mass by consciously bending the knees, using a reverse combat glide.
- Maintain situational awareness of team members, debris, and terrain.
- Use the knees as a shock absorber to steady the body movement to maintain the stability of the upper body, stabilizing the rifle sight(s) on the target.
- Ensure all movement is smooth and steady to maintain stability.
- Bend forward at the waist to put as much mass as possible behind the weapon for recoil management.
- Keep the muzzle oriented downrange toward the expected or detected threat.
- Keep the hips as stationary as possible. Use the upper body as a turret, twisting at the waist, maintaining proper platform with the upper body.

LATERAL MOVEMENT

9-8. Lateral movement is where the Soldier maintains weapon orientation downrange at the expected or detected threat while moving to the left or right. In the most extreme cases, the target will be offset 90 degrees or more from the direction of movement.

TECHNIQUE

9-9. During lateral movement, Soldiers should—

- Place their feet heel to toe and drop their center mass by consciously bending the knees.
- Use the knees as a shock absorber to steady the body movement to maintain the stability of the upper body, stabilizing the rifle sight(s) on the target.
- Ensure all movement is smooth and steady to maintain stability.
- Bend forward at the waist to put as much mass as possible behind the weapon for recoil management.
- Roll the foot, heel to toe, as you place the foot on the ground and lift it up again to provide for the smoothest motion possible.
- Keep the weapon at the alert or ready carry. Do not aim in on the target until ready to engage.
- Maintain awareness of the surroundings, both to the left and right, at all times during movement.
- Trigger control when moving is based on the wobble area. The Soldier shoots when the sights are most stable, not based on foot position.
- Keep the muzzle of the weapon facing down range toward the threat.
- When moving, the placement of the feet should be heel to toe.
- Do not overstep or cross the feet, because this can decrease the Soldier's balance and center of gravity.
- Keep the hips as stationary as possible. Use the upper body as a turret, twisting at the waist, maintaining proper platform with the upper body.

Note. It is more difficult to engage adversaries to the firing side while moving laterally. The twist required to achieve a full 90-degree offset requires proper repetitive training. The basic concept of movement must be maintained, from foot placement to platform.

Twisting at the waist will not allow the weapon to be brought to a full 90 degrees off the direction of travel, especially with nonadjustable butt stocks. The Soldier will need to drop the non-firing shoulder and roll the upper body toward the nonfiring side. This will cause the weapon and upper body to cant at approximately a 45-degree angle, relieving some tension in the abdominal region, allowing the Soldier to gain a few more degrees of offset.

Chapter 9

TURNING MOVEMENT

9-10. Turning movement are used to engage widely dispersed targets in the oblique and on the flanks. Turning skills are just as valuable in a rapidly changing combat environment as firing on the move (such as lateral movement) skills are and should only be used with the alert carry.

9-11. It does not matter which direction the Soldier is turning or which side is the Soldier's strong side. The Soldier must maintain the weapon at an exaggerated low-alert carry for the duration of the turn.

9-12. Muzzle awareness must be maintained at all times. Ensure that the muzzle does not begin to come up on target the body is completely turned toward the threat.

9-13. When executing a turn to either side, the Soldier will—
- **Look first.** Turn head to the direction of the turn first.
- **Weapon follows the eyes.** The Soldier moves the weapon smoothly to where the eyes go.
- **Follow with the body.** The body will begin movement with the movement of the weapon. Soldiers finish the body movement smoothly to maintain the best possible stability for the weapon.
- **Maintain situational awareness.** The Soldier must be completely aware of the surrounding terrain, particularly for tripping hazards. When necessary, Soldiers should visually check their surroundings during the turning action and return their vision to the target area as quickly as possible.

Appendix A

Ammunition

Appendix A discusses the characteristics and capabilities of the different ammunition available for the M4- and M16-series weapons. It also includes general ammunition information such as packaging, standard and North Atlantic Treaty Organization (NATO) marking conventions, the components of ammunition, and general principles of operation. The information within this appendix is 5.56mm for the M4- and M16-series weapons only.

SMALL ARMS AMMUNITION CARTRIDGES

A-1. Ammunition for use in rifles and carbines is described as a cartridge. A small arms cartridge (see figure A-1) is an assembly consisting of a cartridge case, a primer, a quantity of propellant, and a bullet. The following terminology describe the general components of all small arms ammunition (SAA) cartridges:

- **Cartridge case.** The cartridge case is a brass, rimless, center-fire case that provides a means to hold the other components of the cartridge.
- **Propellant.** The propellant (or powder) provides the energy to propel the projectile through the barrel and downrange towards a target through combustion.
- **Primer.** The primer is a small explosive charge that provides an ignition source for the propellant.
- **Bullet.** The bullet or projectile is the only component that travels to the target.

Note. Dummy cartridges are composed of a cartridge case and bullet, with no primer or propellant. Some dummy cartridges contain inert granular materials to simulate the weight and balance of live cartridges.

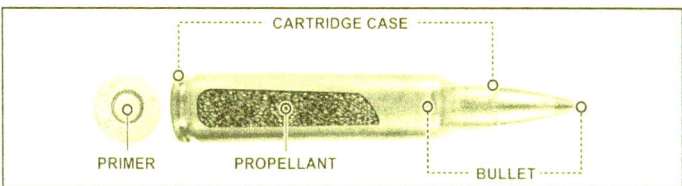

Figure A-1. Small arms ammunition cartridges

Appendix A

A-2. There are multiple types of bullets used for various purposes. These include ball, tracer, armor-piercing, blank, special ball long range (LR), dummy, and short range training.

A-3. The cartridge case is made of steel, aluminum, or a brass combination (70 percent copper and 30 percent zinc) for military use. The M4- and M16-series weapons is a rimless cartridge case that provides an extraction groove (shown in figure A-2). These cartridge cases are designed to support center-fire operation.

A-4. Center-fire cases have a centrally located primer well/pocket in the base of the case, which separates the primer from the propellant in the cartridge case. These cases are designed to withstand pressures generated during firing and are used for most small arms.

A-5. All 5.56mm ammunition uses the rimless cartridge case. A rimless cartridge is where the rim diameter is the same as the case body, and uses an extractor groove to facilitate the cycle of functioning. This design allows for the stacking of multiple cartridges in a magazine.

A-6. When the round is fired, the cartridge case assists in containing the burning propellant by expanding the cartridge case tightly to the chamber walls to provide rear obturation.

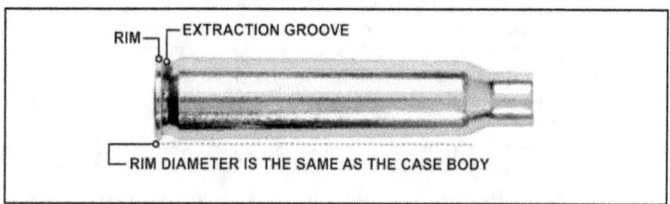

Figure A-2. Cartridge case

PROPELLANT

A-7. Cartridges are loaded with various propellant weights that impart sufficient velocity, within safe pressure, to obtain the required ballistic projectile performance. The propellants are either a single-base (nitrocellulose) or double-base (nitrocellulose and nitroglycerine) composition.

A-8. The propellant (see figure A-3) may be a single-cylindrical or multiple-perforation, a ball, or a flake design to facilitate rapid burning. Most propellants are coated to assist the control of the combustion rate. A final graphite coating facilitates propellant flow and eliminates static electricity in loading the cartridge.

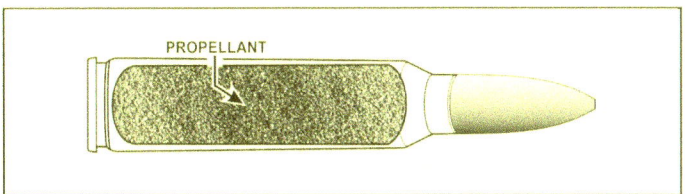

Figure A-3. Propellant

Appendix A

Primer

A-9. Center-fire small arms cartridges contain a percussion primer assembly. The assembly consists of a brass or gilding metal cup (see figure A-4). The cup contains a pellet of sensitive explosive material secured by a paper disk and a brass anvil.

A-10. The weapon firing pin striking the center of the primer cup base compresses the primer composition between the cup and the anvil. This causes the composition to explode. Holes or vents located in the anvil or closure cup allow the flame to pass through the primer vent, igniting the propellant.

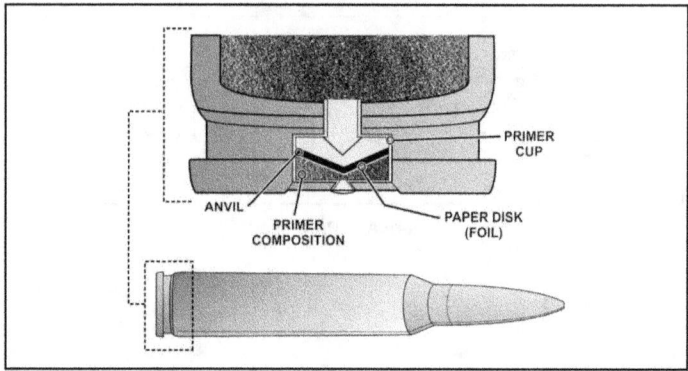

Figure A-4. 5.56mm primer detail

BULLET

A-11. The bullet is a cylindrically shaped lead or alloy projectile that engages with the rifling of the barrel. Newer projectiles consist of a copper slug with exposed steel penetrator, as with the M855A1. The bullets used today are either lead (lead alloy), or assemblies of a jacket and a lead or steel core penetrator. The lead used in lead-alloy bullets is combined with tin, antimony or both for bullet hardness. The alloying reduces barrel leading and helps prevent the bullet from striping (jumping) the rifling during firing.

A-12. Jacketed bullets (see figure A-5) are used to obtain high velocities and are better suited for semiautomatic and automatic weapons. A bullet jacket may be either gilding metal, gilding metal-clad steel, or copper plated steel. In addition to a lead or steel core, they may contain other components or chemicals that provide a terminal ballistic characteristic for the bullet type.

A-13. Some projectiles may be manufactured from plastic, wax, or plastic binder and metal powder, two or more metal powders, or various combinations based on the cartridge's use.

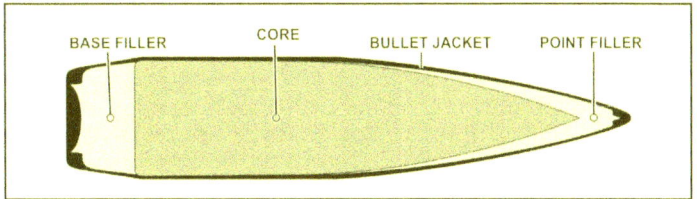

Figure A-5. Bullet example, Armor-piercing cartridge

Appendix A

SMALL ARMS AMMUNITION TYPES

A-14. There are seven types of SAA for the M4- and M16-series weapons that are used for training and combat. Each of these ammunition types provides a different capability and have specific characteristics. The following are the most common types of ammunition for the rifle and carbine:

BALL

A-15. The ball cartridge (see figure A-6) is intended for use in rifles and carbines against personnel and unarmored targets. The bullet, as designed for general purpose combat and training requirements, normally consists of a metal jacket and a lead slug.

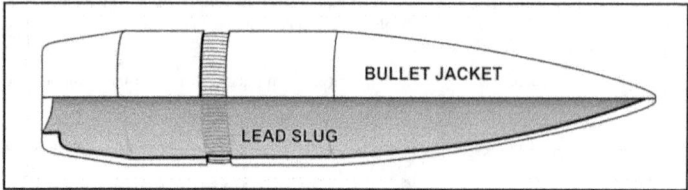

Figure A-6. Ball cartridge

TRACER (TCR OR T)

A-16. A tracer round contains a pyrotechnic composition in the base of the bullet to permit visible observation of the bullet's in-flight path or trajectory and point of impact. (See figure A-7) The pyrotechnic composition is ignited by the propellant when the round is fired, emitting a bright flame visible by the firer. Tracer rounds may also be used to pinpoint enemy targets to ignite flammable materials and for signaling purposes.

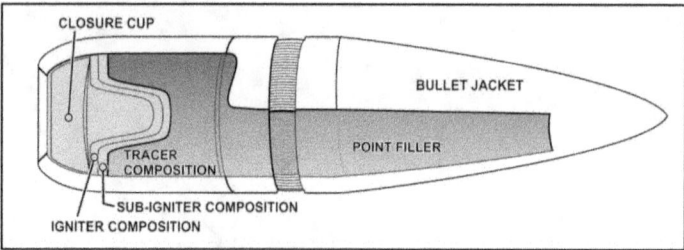

Figure A-7. Ball with tracer cartridge

Ammunition

ARMOR PIERCING (AP)

A-17. The armor-piercing cartridge (see figure A-8) is intended for use against personnel and light armored and unarmored targets, concrete shelters, and similar bullet-resistant targets. The bullet consists of a metal jacket and a hardened steel-alloy core. In addition, it may have a lead base filler and/or a lead point filler.

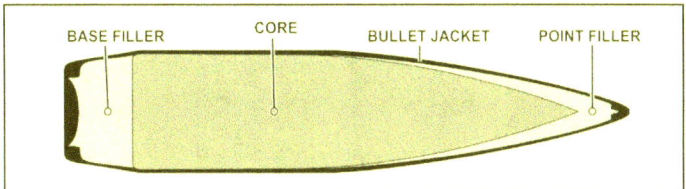

Figure A-8. Armor-piercing cartridge

SHORT RANGE TRAINING AMMUNITION

A-18. The short range training ammunition (SRTA) (see figure A-9) cartridges are designed for target practice where the maximum range is reduced for training purposes. This cartridge ballistically matches the ball cartridge out to 300 meters, and rapidly drops in velocity and accuracy. This allows for installations with restricted training range facilities to continue to operate with accurate munitions. This cartridge is also a preferred round when conducting training in a close quarters environment, like a shoot house or other enclosed training facility.

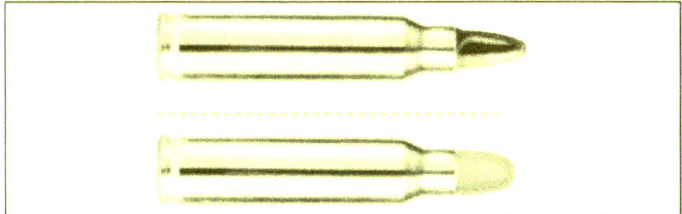

Figure A-9. Short range training ammunition cartridge

Appendix A

BLANK (BLK)

A-19. The blank cartridge (see figure A-10) is distinguished by the absence of a bullet or projectile. It is used for simulated fire, in training maneuvers, and for ceremonial purposes. These rounds consist of a roll crimp (knurl) or cannelure on the body of the case, which holds a paper wad in place instead of a projectile. Newer cartridges have rosette crimp (7 petals) and an identification knurl on the cartridge case.

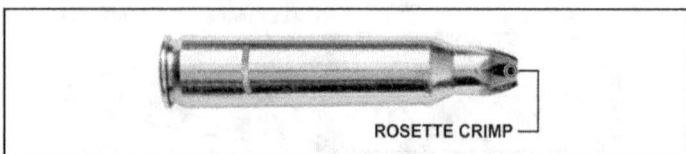

Figure A-10. Blank cartridge

CLOSE COMBAT MISSION CAPABILITY KIT

A-20. The close combat mission capability kit (CCMCK) cartridge (see figure A-11) is used for training purposes only.

A-21. The M4 carbine/M16 rifle conversion adapter kit provides utmost safety, in-service reliability and maintainability. The kit is easy to install with a simple exchange of the bolt. It adapts the host weapon to fire unlinked 5.56mm M1042 man-marking ammunition with the feel and function of live ammunition. The kit includes fail-safe measures to prevent the discharge of a standard "live" round.

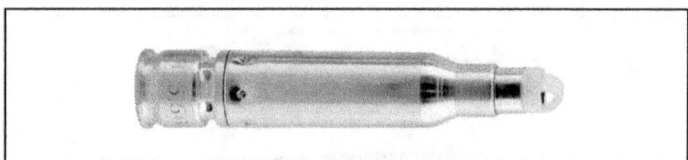

Figure A-11. Close combat mission capability kit cartridge

Ammunition

DUMMY

A-22. The dummy cartridge (see figure A-12) is used for practice in loading weapons and simulated firing to detect errors in employment skills when firing weapons. This round is completely inert and consists only of an empty cartridge case and ball bullet. Cartridge identification is by means of holes through the side of the case or longitudinal corrugations in the case and by the empty primer pocket.

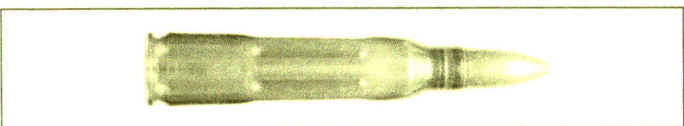

Figure A-12. Dummy cartridge

COLORS, MARKINGS, AND SYMBOLS

A-23. Small arms ammunition is identifiable by color coding specification per type and intended use. Table A-1 describes the general color codes for all types of 5.56mm small arms ammunition. Table A-2 identifies the color code specifications that are applied to the tip of 5.56mm ammunition.

A-24. Markings stenciled or stamped on munitions or their containers include all information needed for complete identification.

A-25. Packaging and containers for small arms ammunition are clearly marked with standard NATO symbols identifying the contents of the package by type of ammunition, primary use, and packaging information. The most common NATO symbols are described according to Standardization Agreement (STANAG) (see table A-2 on page A-11).

A-26. Small arms ammunition (less than 20mm) is not color-coded under MIL-STD-709D. Marking standards for small arms ammunition are outlined in—
- TM 9-1305-201-20&P.
- TM 9-1300-200.

A-27. These publications describe the color coding system for small arms projectiles. The bullet tips are painted a distinctive color as a ready means of identification for the user. (Refer to TM 9-1300-200 for more information.)

Appendix A

Table A-1. Small Arms Color Coding and Packaging Markings

Ammunition Type	Color Coding	Package Marking
Ball	No Color or Green (M855)	●
Tracer (TCR or T)	Orange Tip	■
Armor Piercing (AP)	Black Tip	◀
Short Range Training Ammunition (STRA)	Blue	
Blank (BLK)	Cringed or Capped End	◌
Close Combat Mission Capabilities Kit (CCMCK)	Black Cartridge and Tip, or Perforated Cartridge	None
Dummy		
Special Markings	**Color Code**	**Package Marking**
NATO Standard		⊕
Interchangeable - suitable for use in similar caliber NATO weapons		✠
Bandoleers - ammunition is packaged in bandoleers		▼
Clipped - ammunition is packaged in clips for use with a speed loader		▲▲▲

5.56-MM AMMUNITION

A-28. The following tables A-2 through A-10 on pages A-10 through A-18, will provide a brief description of the ten different types of commonly used 5.56mm ammunition for training and combat. Some types of 5.56mm ammunition will have more than one applicable Department of Defense Identification Code (DODIC); those DODICs are provided for the clarity and ease of the unit's ammunition resource manager.

Table A-2. 5-56mm, M855, Ball

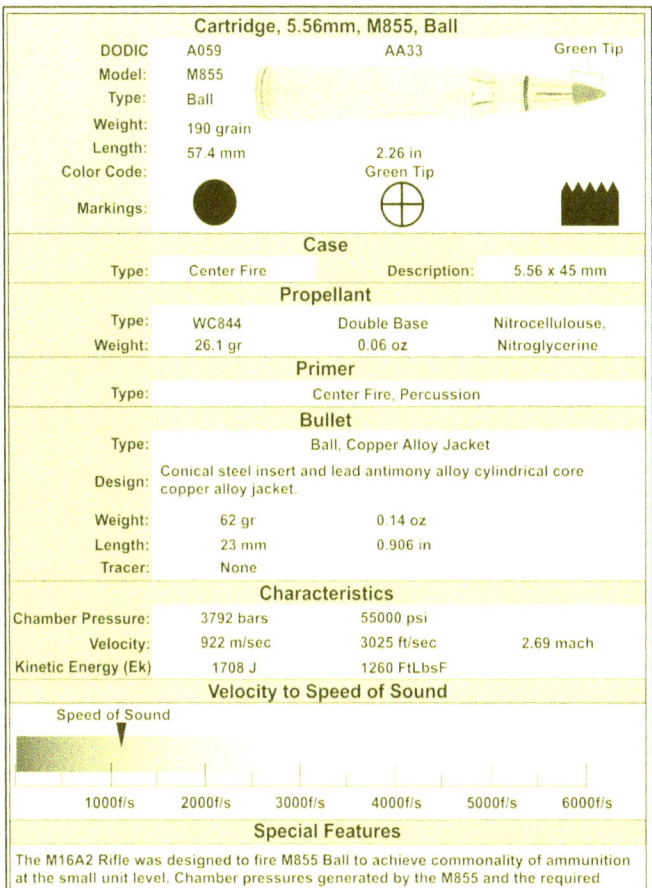

Cartridge, 5.56mm, M855, Ball			
DODIC	A059	AA33	Green Tip
Model:	M855		
Type:	Ball		
Weight:	190 grain		
Length:	57.4 mm	2.26 in	
Color Code:		Green Tip	
Markings:	●	⊕	▲▲▲▲
Case			
Type:	Center Fire	Description:	5.56 x 45 mm
Propellant			
Type:	WC844	Double Base	Nitrocellulouse,
Weight:	26.1 gr	0.06 oz	Nitroglycerine
Primer			
Type:		Center Fire, Percussion	
Bullet			
Type:	Ball, Copper Alloy Jacket		
Design:	Conical steel insert and lead antimony alloy cylindrical core copper alloy jacket.		
Weight:	62 gr	0.14 oz	
Length:	23 mm	0.906 in	
Tracer:	None		
Characteristics			
Chamber Pressure:	3792 bars	55000 psi	
Velocity:	922 m/sec	3025 ft/sec	2.69 mach
Kinetic Energy (Ek)	1708 J	1260 FtLbsF	
Velocity to Speed of Sound			
Speed of Sound ▼ 1000f/s 2000f/s 3000f/s 4000f/s 5000f/s 6000f/s			
Special Features			
The M16A2 Rifle was designed to fire M855 Ball to achieve commonality of ammunition at the small unit level. Chamber pressures generated by the M855 and the required barrel twist (1:7 or 32 calibers) make it unsuitable in the obsolete M16 and M16A1 weapons. The M855's steel insert is effective against most types of fabric body armor while its three-piece construction achieve good effects against unprotected personnel targets.			

Table A-3. 5.56mm, M855A1, Enhanced Performance Round (EPR), Ball

Cartridge, 5.56mm, M855A1, Ball, EPR

DODIC	AB57	AB58	Bronze Tip
Model:	M855A1		
Type:	Ball, EPR		
Weight:	190 grain		
Length:	57.4 mm	2.26 in	
Color Code:		Bronze Tip	
Markings:	●	⊕	▲▲▲

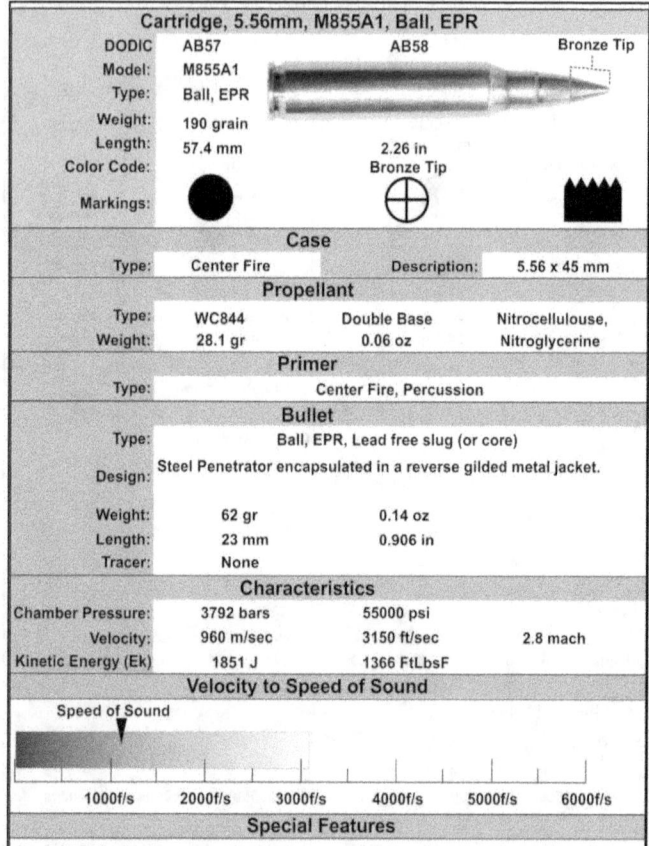

Case

Type:	Center Fire	Description:	5.56 x 45 mm

Propellant

Type:	WC844	Double Base	Nitrocellulouse,
Weight:	28.1 gr	0.06 oz	Nitroglycerine

Primer

Type:	Center Fire, Percussion

Bullet

Type:	Ball, EPR, Lead free slug (or core)
Design:	Steel Penetrator encapsulated in a reverse gilded metal jacket.
Weight:	62 gr — 0.14 oz
Length:	23 mm — 0.906 in
Tracer:	None

Characteristics

Chamber Pressure:	3792 bars	55000 psi	
Velocity:	960 m/sec	3150 ft/sec	2.8 mach
Kinetic Energy (Ek)	1851 J	1366 FtLbsF	

Velocity to Speed of Sound

Speed of Sound ▼

1000f/s   2000f/s   3000f/s   4000f/s   5000f/s   6000f/s

Special Features

The M855A1's steel penetrator is effective against light armored targets while its three-piece construction maintains operational capabilities against unprotected personnel targets. The M855A1 enhances performance on hard targets/barriers. Improved propellant reduces muzzle flash. Optimized for use with the M4 series carbine for close quarters engagements.

Ammunition

Table A-4. 5.56mm, M856A1, Tracer

Cartridge, 5.56mm, M856A1, Tracer			
DODIC	A063		Orange Tip
Model:	M856A1		
Type:	Tracer		
Weight:	190 grain		
Length:	57.4 mm	2.26 in	
Color Code:		Orange Tip	
Markings:			
Case			
Type:	Center Fire	Description:	5.56 x 45 mm
Propellant			
Type:	Wc844	Double Base	Nitrocellulouse,
Weight:	24.7 gr	0.06 oz	Nitroglycerine
Primer			
Type:	Center Fire, Percussion		
Bullet			
Type:	Tracer		
Design:	Lead alloy core in copper alloy jacket with incendiary compound fill in hollow base.		
Weight:	63.7 gr	0.15 oz	
Length:	29.3 mm	1.154 in	
Tracer:	None		
Characteristics			
Chamber Pressure:	3792 bars	55000 psi	
Velocity:	875 m/sec	2870 ft/sec	2.55 mach
Kinetic Energy (Ek)	1580 J	1165 FtLbsF	
Velocity to Speed of Sound			
Speed of Sound — 1000f/s 2000f/s 3000f/s 4000f/s 5000f/s 6000f/s			
Special Features			
Because the M856 loses mass as it travels, it necessitates a 1:7 barrel twist to keep it stable in flight.			

Table A-5. 5.56mm, Mk301, MOD 0, DIM Tracer

Cartridge, 5.56mm, Mk301 Mod 0, Dim Tracer

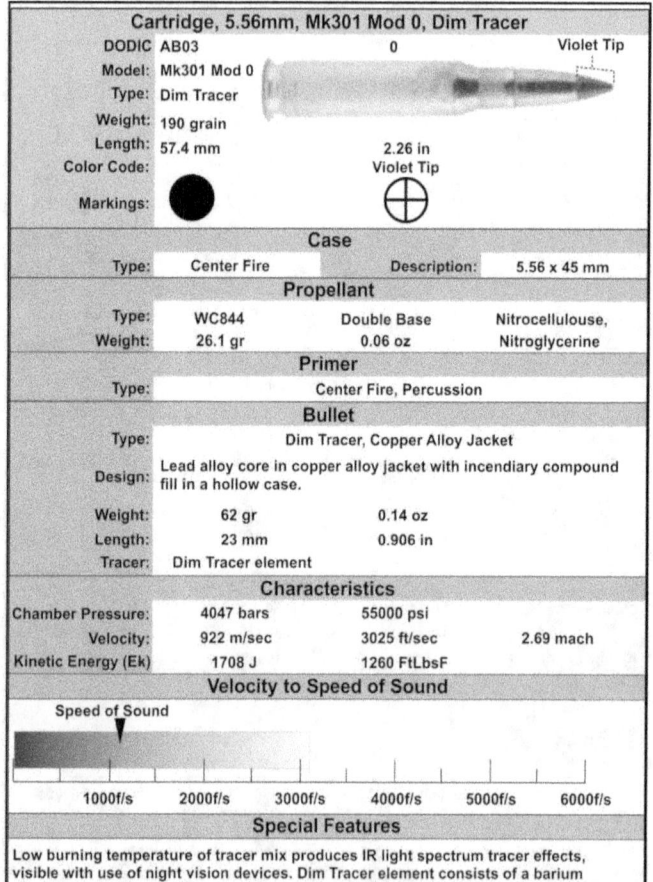

- DODIC: AB03
- Model: Mk301 Mod 0
- Type: Dim Tracer
- Weight: 190 grain
- Length: 57.4 mm / 2.26 in
- Color Code: Violet Tip
- Markings: ● ⊕

Case

Type:	Center Fire	Description:	5.56 x 45 mm

Propellant

Type:	WC844	Double Base	Nitrocellulouse,
Weight:	26.1 gr	0.06 oz	Nitroglycerine

Primer

Type:	Center Fire, Percussion

Bullet

Type:	Dim Tracer, Copper Alloy Jacket
Design:	Lead alloy core in copper alloy jacket with incendiary compound fill in a hollow case.
Weight:	62 gr / 0.14 oz
Length:	23 mm / 0.906 in
Tracer:	Dim Tracer element

Characteristics

Chamber Pressure:	4047 bars	55000 psi	
Velocity:	922 m/sec	3025 ft/sec	2.69 mach
Kinetic Energy (Ek)	1708 J	1260 FtLbsF	

Velocity to Speed of Sound

Speed of Sound ▼

1000f/s 2000f/s 3000f/s 4000f/s 5000f/s 6000f/s

Special Features

Low burning temperature of tracer mix produces IR light spectrum tracer effects, visible with use of night vision devices. Dim Tracer element consists of a barium nitrate composition, with tracer effective range to 900m. WC845S propellant provides flash suppression.

Table A-6. 5.56mm, M995, Armor Piercing

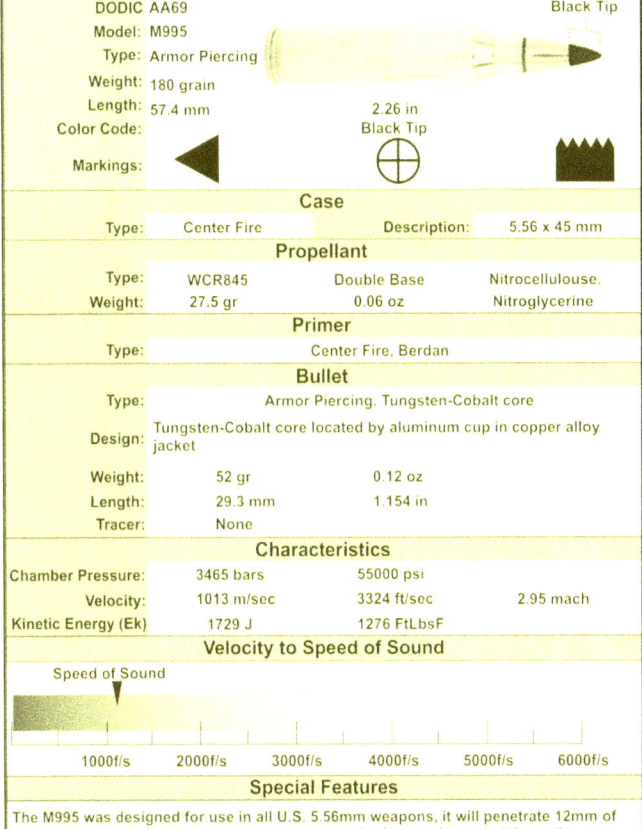

Cartridge, 5.56mm, M995, Armor Piercing			
DODIC:	AA69		Black Tip
Model:	M995		
Type:	Armor Piercing		
Weight:	180 grain		
Length:	57.4 mm	2.26 in	
Color Code:		Black Tip	
Markings:			
Case			
Type:	Center Fire	Description:	5.56 x 45 mm
Propellant			
Type:	WCR845	Double Base	Nitrocellulouse,
Weight:	27.5 gr	0.06 oz	Nitroglycerine
Primer			
Type:		Center Fire, Berdan	
Bullet			
Type:	Armor Piercing, Tungsten-Cobalt core		
Design:	Tungsten-Cobalt core located by aluminum cup in copper alloy jacket		
Weight:	52 gr	0.12 oz	
Length:	29.3 mm	1.154 in	
Tracer:	None		
Characteristics			
Chamber Pressure:	3465 bars	55000 psi	
Velocity:	1013 m/sec	3324 ft/sec	2.95 mach
Kinetic Energy (Ek)	1729 J	1276 FtLbsF	
Velocity to Speed of Sound			
Speed of Sound — 1000f/s 2000f/s 3000f/s 4000f/s 5000f/s 6000f/s			
Special Features			
The M995 was designed for use in all U.S. 5.56mm weapons, it will penetrate 12mm of steel at 100m to defeat light armored vehicles and other barrier materials on the battlefield.			

Table A-7. 5.56mm, M862, Short Range Training Ammunition

Cartridge, 5.56mm, M862, SRTA			
DODIC	AA68		Blue Tip
Model:	M862		
Type:	SRTA		
Weight:	108 grain		
Length:	57.4 mm	2.26 in	
Color Code:		Fluted blue Tip	
Markings:			

Case			
Type:	Center Fire	Description:	5.56 x 45 mm

Propellant			
Type:	0	Double Base	Nitrocellulouse,
Weight:	gr	0 oz	Nitroglycerine

Primer	
Type:	Center Fire

Bullet		
Type:		SRTA
Design:	Plastic projectile	
Weight:	6.9 gr	0.02 oz
Length:	29.3 mm	1.154 in
Tracer:	None	

Characteristics			
Chamber Pressure:	2758 bars	40000 psi	
Velocity:	1379 m/sec	4525 ft/sec	4.02 mach
Kinetic Energy (Ek)	425 J	314 FtLbsF	

Velocity to Speed of Sound

Speed of Sound
1000f/s 2000f/s 3000f/s 4000f/s 5000f/s 6000f/s

Special Features

The M862 is ballistically matched to standard M855 ball ammunition out to 25m, with a maximum range of 250m. M862 ammunition MUST be used with the M2 training bolt. This provides units the capability to conduct training on installations that have limited range facilities which require the use of reduced/decreased Surface Danger Zones.

Table A-8. 5.56mm, M1037, Short Range Training Ammunition

Cartridge, 5.56mm, M1037, SRTA			
DODIC	AB67		Blue Tip
Model:	M1037		
Type:	SRTA		
Weight:	165 grain		
Length:	57.4 mm	2.26 in	
Color Code:		Fluted blue Tip	
Markings:			
Case			
Type:	Center Fire	Description:	5.56 x 45 mm
Propellant			
Type:	0	Double Base	Nitrocellulouse,
Weight:	gr	0 oz	Nitroglycerine
Primer			
Type:		Center Fire	
Bullet			
Type:		SRTA, Frangible	
Design:	Copper, nylon and carbon fiber projectile		
Weight:	33 gr	0.08 oz	
Length:	29.3 mm	1.154 in	
Tracer:	None		
Characteristics			
Chamber Pressure:	2758 bars	40000 psi	
Velocity:	1097 m/sec	3600ft/sec	3.2 mach
Kinetic Energy (Ek)	1287J	950 FtLbsF	
Velocity to Speed of Sound			
Speed of Sound			
1000f/s	2000f/s	3000f/s 4000f/s	5000f/s 6000f/s
Special Features			
The M1037 is ballistically matched to standard M855 ball ammunition out to 100m, with a maximum range of less than 600m. M1037 ammunition DOES NOT require the use of the M2 training bolt. This provides units the capability to conduct training on installations that have limited range facilities which require the use of reduced/decreased Surface Danger Zones.			

Appendix A

Table A-9. 5.56mm, M1042 Close Combat Mission Capability Kit

Cartridge, 5.56mm, M1042, Close Combat Mission Capability Kit			
DODIC	AB09 (blue tip)	AB10 (red tip)	AB11 (yellow tip)
Model:	M1042		
Type:	CCMCK	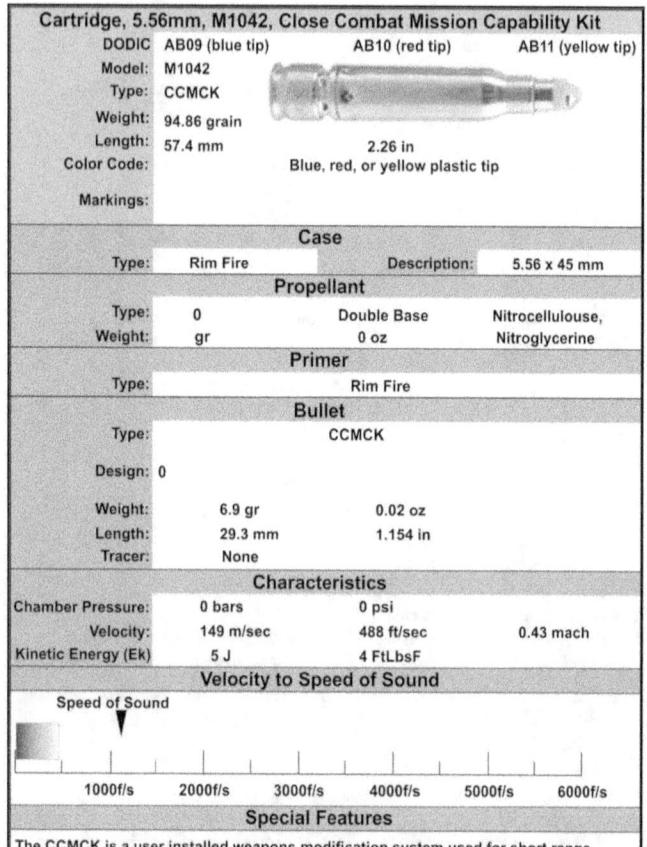	
Weight:	94.86 grain		
Length:	57.4 mm	2.26 in	
Color Code:		Blue, red, or yellow plastic tip	
Markings:			
Case			
Type:	Rim Fire	Description:	5.56 x 45 mm
Propellant			
Type:	0	Double Base	Nitrocellulouse,
Weight:	gr	0 oz	Nitroglycerine
Primer			
Type:		Rim Fire	
Bullet			
Type:		CCMCK	
Design: 0			
Weight:	6.9 gr	0.02 oz	
Length:	29.3 mm	1.154 in	
Tracer:	None		
Characteristics			
Chamber Pressure:	0 bars	0 psi	
Velocity:	149 m/sec	488 ft/sec	0.43 mach
Kinetic Energy (Ek)	5 J	4 FtLbsF	
Velocity to Speed of Sound			
Speed of Sound — 1000f/s 2000f/s 3000f/s 4000f/s 5000f/s 6000f/s			
Special Features			
The CCMCK is a user installed weapons modification system used for short range force on force training. The M1042 is a low velocity marking ammunition that prevents the weapon from firing service ammunition. Fail-safe is achieved by utilizing a 3mm offset firing pin which will only work with the M1042 rim fire primer. In the event that a "Live" 5.56mm cartridge is chambered and the trigger is pulled, the conversion will offset.			

Table A-10. 5.56mm, M200, Blank

Cartridge, 5.56mm, M200, Blank			
DODIC:	A080		Rosette Crimp
Model:	M200		
Type:	Blank		
Weight:	107 grain		
Length:	48.3 mm	1.902in	
Color Code:		Rosette Crimp	
Markings:			

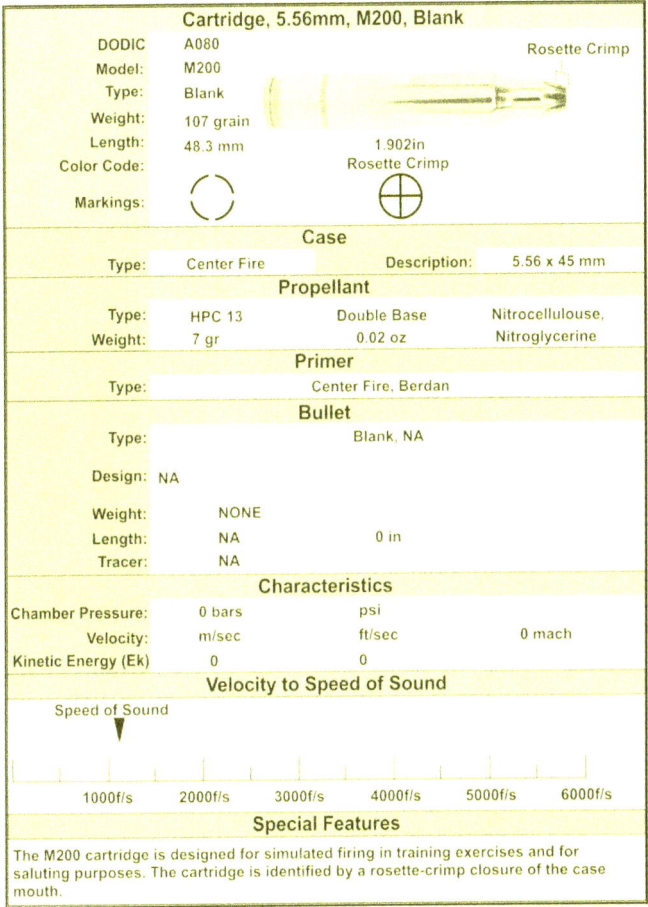

Case			
Type:	Center Fire	Description:	5.56 x 45 mm

Propellant			
Type:	HPC 13	Double Base	Nitrocellulouse,
Weight:	7 gr	0.02 oz	Nitroglycerine

Primer	
Type:	Center Fire, Berdan

Bullet	
Type:	Blank, NA
Design:	NA
Weight:	NONE
Length:	NA 0 in
Tracer:	NA

Characteristics			
Chamber Pressure:	0 bars	psi	
Velocity:	m/sec	ft/sec	0 mach
Kinetic Energy (Ek)	0	0	

Velocity to Speed of Sound

Speed of Sound ▼

1000f/s   2000f/s   3000f/s   4000f/s   5000f/s   6000f/s

Special Features

The M200 cartridge is designed for simulated firing in training exercises and for saluting purposes. The cartridge is identified by a rosette-crimp closure of the case mouth.

This page intentionally left blank.

Appendix B
Ballistics

Ballistics is the science of the processes that occur from the time a firearm is fired to the time when the bullet impacts its target. Soldiers must be familiar with the principles of ballistics as they are critical in understanding how the projectiles function, perform during flight, and the actions of the bullet when it strikes the intended target. The profession of arms requires Soldiers to understand their weapons, how they operate, their functioning, and their employment.

B-1. The flight path of a bullet includes three stages: the travel down the barrel, the path through the air to the target, and the actions the bullet takes upon impact with the target. These stages are defined in separate categories of ballistics: internal, external, and terminal ballistics.

INTERNAL BALLISTICS

B-2. **Internal ballistics** – is the study of the propulsion of a projectile. Internal ballistics begin from the time the firing pin strikes the primer to the time the bullet leaves the muzzle. Once the primer is struck the priming charge ignites the propellant. The expanding gases caused by the burning propellant create pressures which push the bullet down the barrel. The bullet engages the lands and grooves (rifling) imparting a spin on the bullet that facilitates stabilization of the projectile during flight. Internal ballistics ends at shot exit, where the bullet leaves the muzzle. (See figure B-1.)

Appendix B

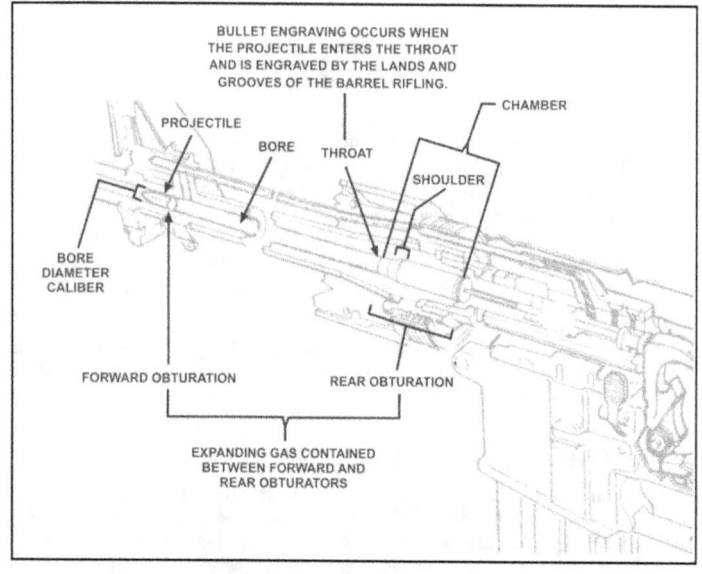

Figure B-1. Internal ballistic terms

B-3. Several key terms are used when discussing the physical actions of internal ballistics —

- **Bore** – the interior portion of the barrel forward of the chamber.
- **Chamber** – the part of the barrel that accepts the ammunition for firing.
- **Grain (gr)** – a unit of measurement of either a bullet or a projectile. There are 7000 grains in a pound, or 437.5 grains per ounce.
- **Pressure** – the force developed by the expanding gasses generated by the combustion (burning) of the propellant. Pressure is measure in pounds per square inch (psi).
- **Shoulder** – the area of the chamber that contains the shoulder, forcing the cartridge and projectile into the entrance of the bore at the throat of the barrel.
- **Muzzle** – the end of the barrel.
- **Throat** – the entrance to the barrel from the chamber. Where the projectile is introduced to the lands and grooves within the barrel.

Ballistics

EXTERNAL BALLISTICS

B-4. **External ballistics** is the study of the physical actions and effects of gravity, drag, and wind along the projectile's flight to the target. It includes only those general physical actions that cause the greatest change to the flight of a projectile. (See figure B-2.) External ballistics begins at shot exit and continues through the moment the projectile strikes the target.

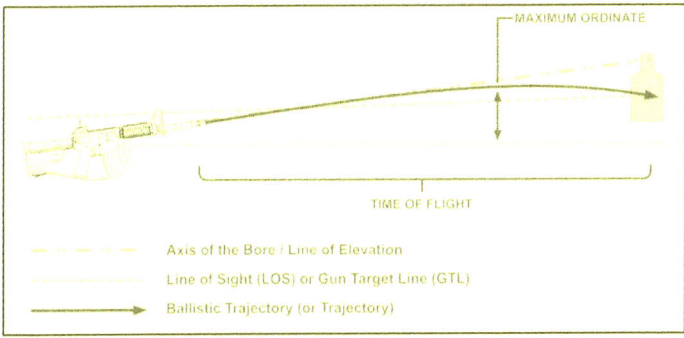

Figure B-2. External ballistic terms

B-5. The following terms and definitions are used to describe the actions or reactions of the projectile during flight. This terminology is standard when dealing with any weapon or weapon system, regardless of caliber. (See figure B-3.)

- **Axis of the bore** (Line of Bore) – the line passing through the center of the bore or barrel.
- **Line of sight (LOS) or gun target line (GTL)** – a straight line between the sights or optics and the target. This is never the same as the axis of the bore. The LOS is what the Soldier sees through the sights and can be illustrated by drawing an imaginary line from the firer's eye through the rear and front sights out to infinity. The LOS is synonymous with the GTL when viewing the relationship of the sights to a target.
- **Line of elevation (LE)** – the angle represented from the ground to the axis of the bore.
- **Ballistic trajectory** – the path of a projectile when influenced only by external forces, such as gravity and atmospheric friction.
- **Maximum ordinate** – the maximum height the projectile will travel above the line of sight on its path to the point of impact.
- **Time of flight** – the time taken for a specific projectile to reach a given distance after firing.

Appendix B

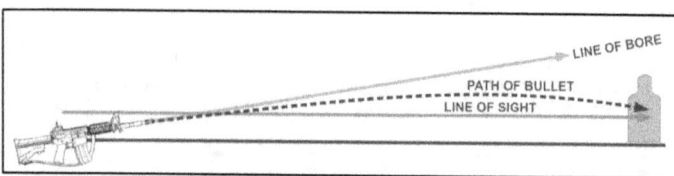

Figure B-3. Trajectory

- **Jump** – vertical jump in an upward and rearward direction caused by recoil. Typically, it is the angle, measured in mils, between the line of departure and the line of elevation.
- **Line of departure (LD)** – the line the projectile is on at shot exit.
- **Muzzle** – the end of the barrel.
- **Muzzle velocity or velocity** – the velocity of the projectile measured at shot exit. Muzzle velocity decreases over time due to air resistance. For small arms ammunition, velocity (V) is represented in feet per second (f/s).
- **Twist rate** – the rotation of the projectile within the barrel of a rifled weapon based on the distance to complete one revolution. The twist rate relates to the ability to gyroscopically spin-stabilize a projectile on rifled barrels, improving its aerodynamic stability and accuracy. The twist rate of the M4- or M16-series weapon is a right hand, one revolution in every seven inches of barrel length (or R 1:7 inches).
- **Shot exit** – the moment the projectile clears the muzzle of the barrel, where the bullet is not supported by the barrel.
- **Oscillation** – the movement of the projectile in a circular pattern around its axis during flight.
- **Drift** – the lateral movement of a projectile during its flight caused by its rotation or spin.
- **Yaw** – a deviation from stable flight by oscillation. This can be caused by cross wind or destabilization when the projectile enters or exits a transonic stage.
- **Grain (gr)** – a unit of measurement of either a bullet or a propellant charge. There are 7000 grains in a pound, or 437.5 grains per ounce.
- **Pressure** – the force developed by the expanding gases generated by the combustion (burning) of the propellant. For small arms, pressure is measured in pounds per square inch (psi).
- **Gravity** – the constant pressure of the earth on a projectile at a rate of about 9.8 meters per second squared, regardless of the projectile's weight, shape or velocity. Commonly referred to as bullet drop, gravity causes the projectile to drop from the line of departure. Soldiers must understand the effects of gravity on the projectile when zeroing as well as how it applies to determining the appropriate hold-off at ranges beyond the zero distance.

Ballistics

- **Drag (air resistance)** – the friction that slows the projectile down while moving through the air. Drag begins immediately upon the projectile exiting the barrel (shot exit). It slows the projectile's velocity over time, and is most pronounced at extended ranges. Each round has a ballistic coefficient (BC) that is a measurement of the projectile's ability to minimize the effects of air resistance (drag) during flight.
- **Trajectory** – the path of flight that the projectile takes upon shot exit over time. For the purposes of this manual, the trajectory ends at the point of impact.
- **Wind** – has the greatest variable effect on ballistic trajectories. The effects of wind on a projectile are most noticeable in three key areas between half and two-thirds the distance to the target:
  - **Time (T)** – the amount of time the projectile is exposed to the wind along the trajectory. The greater the range to target, the greater time the projectile is exposed to the wind's effects.
  - **Direction** – the direction of the wind in relation to the axis of the bore. This determines the direction of drift of the projectile that should be compensated.
  - **Velocity (V)** – the speed of the wind during the projectile's trajectory to the target. Variables in the overall wind velocity affecting a change to the ballistic trajectory include sustained rate of the wind and gust spikes in velocity.

TERMINAL BALLISTICS

B-6. Terminal ballistics is the science of the actions of a projectile from the time it strikes an object until it comes to rest (called terminal rest). This includes the terminal effects that take place against the target.

- **Kinetic Energy (E_K)** – a unit of measurement of the delivered force of a projectile. Kinetic energy is the delivered energy that a projectile possesses due to its mass and velocity at the time of impact. Kinetic energy is directly related to the *penetration capability* of a projectile against the target.
- **Penetration** – the ability or act of a projectile to enter a target's mass based on its delivered kinetic energy. When a projectile strikes a target, the level of penetration into the target is termed the impact depth. The impact depth is the distance from the point of impact to the moment the projectile stops at its terminal resting place. Ultimately, the projectile stops when it has transferred its momentum to an equal mass of the medium (or arresting medium).

B-7. Against any target, penetration is the most important terminal ballistic consideration. Soldiers must be aware of the penetration capabilities of their ammunition against their target, and the most probable results of the terminal ballistics.

B-8. The 5.56mm projectile's purpose is to focus the largest amount of momentum (energy) on the smallest possible area of the target to achieve the greatest penetration. They are designed to resist deformation on impact to enter the target's mass. The steel

Appendix B

tip of the penetrator allows for reduced deformation through light skin armor or body armor, and the heavier steel penetrator allows for increased soft tissue damage.

ACTIONS AFTER THE TRIGGER SQUEEZE

B-9. Once the trigger is squeezed, the ballistic actions begin. Although not all ammunition and weapons operate in the same manner, the following list describes the general events that occur on the M4- and M16-series weapons when the trigger is squeezed.

- The hammer strikes the rear of the firing pin.
- The firing pin is pushed forward, striking the cartridge percussion primer assembly.
- The primer is crushed, pushing the primer composition through the paper disk, and on to the anvil, detonating the primer composition.
- The burning primer composition is focused evenly through the primer cup vent hole, igniting the propellant.
- The propellant burns evenly within the cartridge case.
- The cartridge case wall expand from the pressure of the burning propellant, firmly locking the case to the chamber walls.
- The expanded cartridge case, held firmly in place by the chamber walls and the face of the bolt provide rear obturation, keeping the burning propellant and created expanding gasses in front of the cartridge case.
- The projectile is forced by the expanding gasses firmly into the lands and grooves at the throat of the bore, causing engraving.
- Engraving causes the scoring of the softer outer jacket of the projectile with the lands and grooves of the bore. This allows the projectile to spin at the twist rate of the lands and grooves, and provides a forward obturation seal. The forward obturation keeps the expanding gasses behind the projectile in order to push it down the length of the barrel.
- As the propellant continues to burn, the gasses created continue to seek the path of least resistance. As the cartridge case is firmly seated and the projectile is moveable, the gas continues to exert its force on the projectile.
- Once the projectile passes the gas port on the top of the barrel, a small amount of gas is permitted to escape from propelling the projectile. This escaping gas is directed up through the gas port and rearward through the gas tube, following the path of least resistance. The diameter of the gas port limits the amount of gas allowed to escape.
- As the end of the projectile leaves the muzzle, it is no longer supported by the barrel itself. Shot exit occurs.
- Upon shot exit, most of the expanding and burning gasses move outward and around the projectile, causing the muzzle flash.
- At shot exit, the projectile achieves its maximum muzzle velocity. From shot exit until the projectile impacts an object, the projectile loses velocity at a steady rate due to air resistance.

Ballistics

- As the round travels along its trajectory, the bullet drops consistently by the effects of gravity.
- As the actual line of departure is an elevated angle from the line of sight, the projectile appears to rise and then descend. This rise and fall of the projectile is the trajectory.
- The round achieves the highest point of its trajectory typically over half way to the target, depending on the range to target. The high point is called the round's maximum ordinate or *max ord*.
- From the max ord, the projectile descends into the target.
- The round strikes the target at the point of impact, which, depending on the firing event, may or may not be the desired point of impact, and is seldom the point of aim.

Note. The point of aim and point of impact only occur twice during the bullet's path to the target at distance: once when the trajectory crosses the line of sight approximately 25 meters from the muzzle, and again at the zero distance (300 meters for the Army standard zero).

- Once the projectile strikes a target or object, it delivers its kinetic energy (force) at the point of impact.
- Terminal ballistics begin.

B-10. Once terminal ballistics begin, no bullets follow the same path or function. Generally speaking, the projectile will penetrate objects where the delivered energy (mass times velocity squared, divided by 2) is greater than the mass, density, and area of the target at the point of the delivered force. There are other contributing factors, such as the angle of attack, yaw, oscillation, and other physical considerations that are not included in this ballistic discussion.

STRUCTURE PENETRATION

B-11. The following common barriers in built-up areas can prevent penetration by a 5.56-mm round fired at less than 50 meters (M855) including:
- Single row sandbags.
- A 2-inch thick concrete wall (not reinforced with rebar or similar item).
- A 55-gallon drum filled with water or sand.
- A metal ammunition can filled with sand.
- A cinder block filled with sand (the block may shatter).
- A plate glass windowpane at a 45-degree angle (glass fragments will be thrown behind the glass).
- A brick veneer.

Note. The M855A1 enhanced performance round (EPR) has increased capabilities for barrier penetration compared with M855 as shown above.

Appendix B

B-12. Although most structural materials repel single 5.56-mm rounds, continued and concentrated firing can breach (penetrate through) some typical urban structures.

B-13. The best method for breaching a masonry wall is by firing short bursts in a U-shaped pattern. The distance from the firer to the wall should be minimized for best results—ranges as close as 25 meters are relatively safe from ricochet.

B-14. Ball ammunition and armor-piercing rounds produce almost the same results, but are more likely to ricochet to the sides and rearward back at the firer (called spit-back).

> *Note.* Soldiers must ensure the appropriate level of personal protective equipment is worn when conducting tactical and collective tasks, particularly at ranges less than 50 meters.

B-15. The 5.56-mm round can be used to create either a loophole (about 7 inches in diameter) or a breach hole (large enough for a man to enter). When used against reinforced concrete, the M16 rifle and M249 cannot cut the reinforcing bars.

SOFT TISSUE PENETRATION

B-16. A gunshot wound, or ballistic trauma, is a form of physical damage sustained from the entry of a projectile. The degree of tissue disruption caused by a projectile is related to the size of the cavities created by the projectile as it passes through the target's tissue. When striking a personnel target, there are two types of cavities created by the projectile: permanent and temporary wound cavities.

Permanent Wound Cavity

B-17. The permanent cavity refers specifically to the physical hole left in the tissues of soft targets by the pass-through of a projectile. It is the total volume of tissue crushed or destroyed along the path of the projectile within the soft target.

B-18. Depending on the soft tissue composition and density, the tissues are either elastic or rigid. Elastic organs stretch when penetrated, leaving a smaller wound cavity. Organs that contain dense tissue, water, or blood are rigid, and can shatter from the force of the projectile. When a rigid organ shatters from a penetrating bullet, it causes massive blood loss within a larger permanent wound cavity. Although typically fatal, striking these organs may not immediately incapacitate the target.

Temporary Wound Cavity

B-19. The temporary wound cavity is an area that surrounds the permanent wound cavity. It is created by soft, elastic tissues as the projectile passes through the tissue at greater than 2000 feet per second. The tissue around the permanent cavity is propelled outward (stretched) in an almost explosive manner from the path of the bullet. This forms a temporary recess or cavity 10 to 12 times the bullet's diameter.

B-20. Tissue such as muscle, some organs, and blood vessels are very elastic and can be stretched by the temporary cavity with little or no damage and have a tendency to absorb the projectile's energy. The temporary cavity created will slowly reduce in size

Ballistics

over time, although typically not returning completely to the original position or location.

> *Note.* Projectiles that do not exceed 2000 feet per second velocity on impact do not provide sufficient force to cause a temporary cavity capable of incapacitating a threat.

B-21. The extent of the cavitation (the bullet's creation of the permanent and temporary cavities) is related to the characteristics of the projectile:

- **Kinetic energy (E_K)** – the delivered mass at a given velocity. Higher delivered kinetic energy produces greater penetration and tissue damage.
- **Yaw** – any yaw at the point of impact increases the projectiles surface area that strikes the target, decreasing kinetic energy, but increasing the penetration and cavity size.
- **Deformation** – the physical changes of the projectile's original shape and design due to the impact of the target. This increases the projectile's surface area and the size of the cavity created after penetration.
- **Fragmentation** – the fracturing of a projectile into multiple pieces or sub-projectiles. The multiple paths of the fragmented sub-projectiles are unpredictable in size, velocity, and direction. The bullet jacket, and for some types of projectiles, the lead core, fracture creating small, jagged, sharp edged pieces that are propelled outward with the temporary cavity. Fragments can sever tissue, causing large, seemingly explosive-type. Bone fragments caused by the bullet's strike can have the same effect.
- **Tumbling** – the inadvertent end-over-end rotation of the projectile. As a projectile tumbles as it strikes the target, the bullet travels through the tissues with a larger diameter. This causes a more severe permanent cavity as it passes through the soft tissue. A tumbling projectile can change direction erratically within the body due to its velocity and tendency to strike dense material with a larger surface area.

B-22. Once inside the target, the projectile's purpose is to destroy soft tissues with fragmentation. The ball ammunition is designed to not flatten or expand on impact, which would decrease velocity and delivered energy. For the M855-series cartridge, the penetrator tends to bend at the steel-core junction, fracture the weaker jacketed layer, and fragment into pieces when striking an object.

Incapacitation

B-23. Incapacitation with direct fire is the act of ballistically depriving a target of the ability, strength, or capability to continue its tactical mission. To assist in achieving the highest probability of incapacitation with a single shot, the projectile is designed with the ability to tumble, ricochet, or fragment after impact.

B-24. The projectile or its fragments then must hit a vital, blood-bearing organ or the central nervous system to effectively incapacitate the threat. The projectile's limited fragmentation potential after entry maximizes the soft tissue damage and increases the potential for rapid incapacitation.

Lethal Zones

B-25. The Soldier's primary point of aim at any target by default is center of visible mass. This allows for a tolerance that includes the greatest margin of error with the highest probability of a first round hit. The combat conditions may require more precise fires at partially exposed targets or targets that require immediate incapacitation.

B-26. Ideally, the point of aim is anywhere within a primary switch area. This point will maximize the possibility of striking major organs and vessels, rendering a clean, one-shot kill (see figure B-4.)

Ballistics

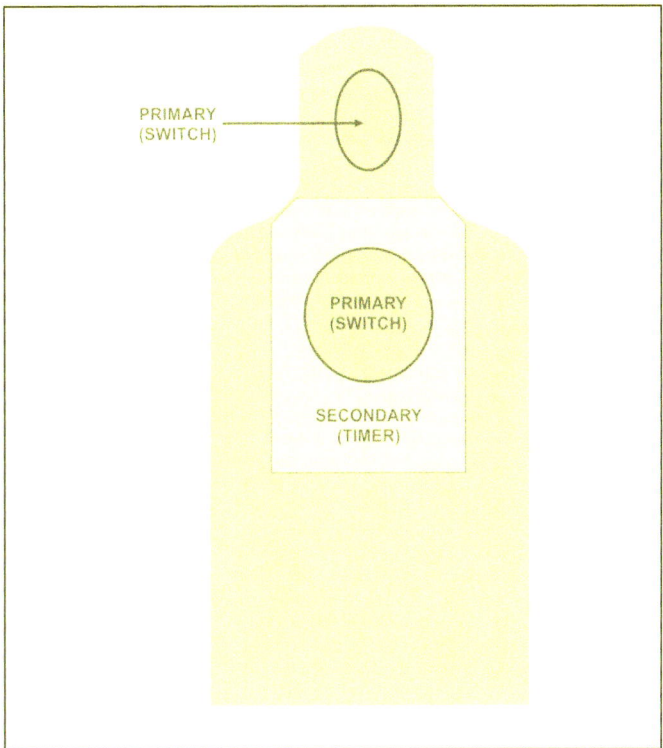

Figure B-4. Lethal zone example

B-27. Shots to the head should be weighed with caution. The head is the most frequently moved body part and are the most difficult to hit with precision. Shots to other exposed body parts, such as the pelvic area, should be considered for the shot.

B-28. Shots to the pelvic area are used when the target is not completely visible or when the target is wearing body armor that prevents the Soldier from engaging the primary zone. This area is rich in large blood vessels and a shot here has a good possibility of impeding enemy movement by destroying the pelvic or hitting the lower spine.

- Circuitry shots (switches).
- Hydraulic shots (timers).

Circuitry Shots (Switches)

B-29. Circuitry shots, or "switches," are strikes to a target that deliver its immediate incapacitation. Immediate incapacitation is the sudden physical or mental inability to initiate or complete any physical task. To accomplish this, the central nervous system must be destroyed by hitting the brain or spinal column. All bodily functions and voluntary actions cease when the brain is destroyed and if the spinal column is broken, all functions cease below the break.

Hydraulic Shots (Timer)

B-30. Hydraulic shots, or "timers," are impacts on a target where immediate incapacitation is not guaranteed. These types of ballistic trauma are termed "timers" as that after the strike of the bullet, the damage caused requires time for the threat to have sufficient blood loss to render it incapacitated. Hydraulic shots, although ultimately lethal, allow for the threat to function in a reduced capacity for a period of time.

B-31. For hydraulic shots to eliminate the threat, they must cause a 40 percent loss of blood within the circulatory system. If the shots do not disrupt that flow at a rapid pace, the target will be able to continue its mission. Once two (2) liters of blood are lost, the target will transition into hypovolemic shock and become incapacitated.

Appendix C
Complex Engagements

*This appendix provides detailed information on the calculations for determining **deliberate** holds for complex engagements and various engagement techniques. It is designed for the advanced shooter; however, all Soldiers should be familiar with the contents of the appendix in order to build their mastery and proficiency with their individual weapon.*

C-1. A complex engagement includes any shot that cannot use the *CoVM* as the point of aim to ensure a target hit. Complex engagements require a Soldier to apply various points of aim (called hold, hold-off, or holds) to successfully defeat the threat.

C-2. This appendix builds upon the concepts discussed in Chapter 7, Aim, and only include topics specific to deliberate hold determinations. These topics are:
- **Target conditions**:
  - Range to target.
  - Moving targets.
  - Oblique targets.
- **Environmental conditions**:
  - Wind.
  - Angled firing.
- **Compound conditions**:

C-3. Each of these firing conditions may require the Soldier to determine an appropriate aim point that is not the CoVM. During any complex engagement, the Soldier serves as the ballistic computer during the shot process. The hold represents a refinement or alteration of the center of visible mass point of aim at the target to counteract certain conditions during a complex engagement for—
- Range to target.
- Lead for targets based on their direction and speed of movement.
- Counter-rotation lead required when the Soldier is moving in the opposite direction of the moving target.
- Wind speed, direction, and duration between the shooter and the target at ranges greater than 300 meters.
- Greatest lethal zone presented by the target to provide the most probable point of impact to achieve immediate incapacitation.

C-4. The Soldier will apply the appropriate aim (hold) based on the firing instances presented. Hold determinations will be discussed in two formats; immediate and deliberate.

Appendix C

TARGET CONDITIONS

C-5. Soldiers must consider several aspects of the target to apply the proper point of aim on the target. The target's posture, or how it is presenting itself to the shooter, consists of—
- Range to target.
- Nature of the target.
- Nature of the terrain (surrounding the target).

RANGE TO TARGET

C-6. Rapidly determining an accurate range to target is critical to the success of the Soldier at mid and extended ranges. There are several range determination methods shooters should be confident in applying to determine the proper hold-off for pending engagements.

Deliberate Range Determination

C-7. The deliberate methods afford the shooter a reliable means of determining the range to a given target; however, these methods require additional time. (See figure C-1.) With practice and experience, the time to determine the range with these methods is reduced significantly. The various methods of deliberate range determination are:
- Reticle relationship (mil or MOA).
- Recognition method.
- Bracketing method.
- Halving method.

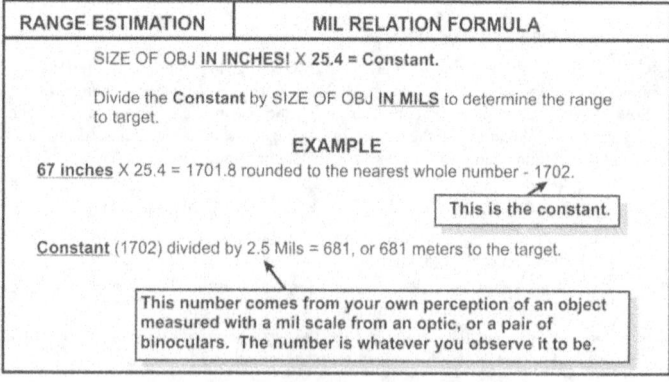

Figure C-1. Mil Relation Formula example

Complex Engagements

Reticle Relationship Method

C-8. With this method, shooters use their aiming device's reticle to determine the range to target based on standard target information. To use the appearance of objects method based on how they align to an aiming device's reticle, shooters must be familiar with the sizes and details of personnel and equipment at known distances as shown in figure C-2.

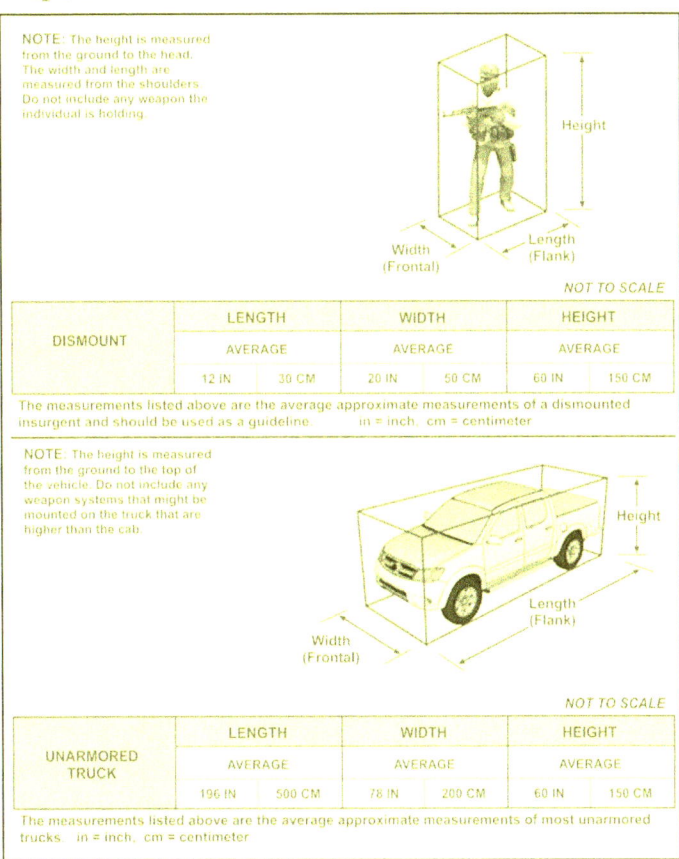

Figure C-2. Standard dismount threat dimensions example

Appendix C

C-9. Knowing the standard dimensions to potential targets allows for the Soldier to assess those dimensions using the aiming device's reticle. The Soldier will apply the mil or MOA relationship as they pertain to the aiming device and the target. Figure C-3 and figure C-4 on page C-5, show various reticle relationship examples.

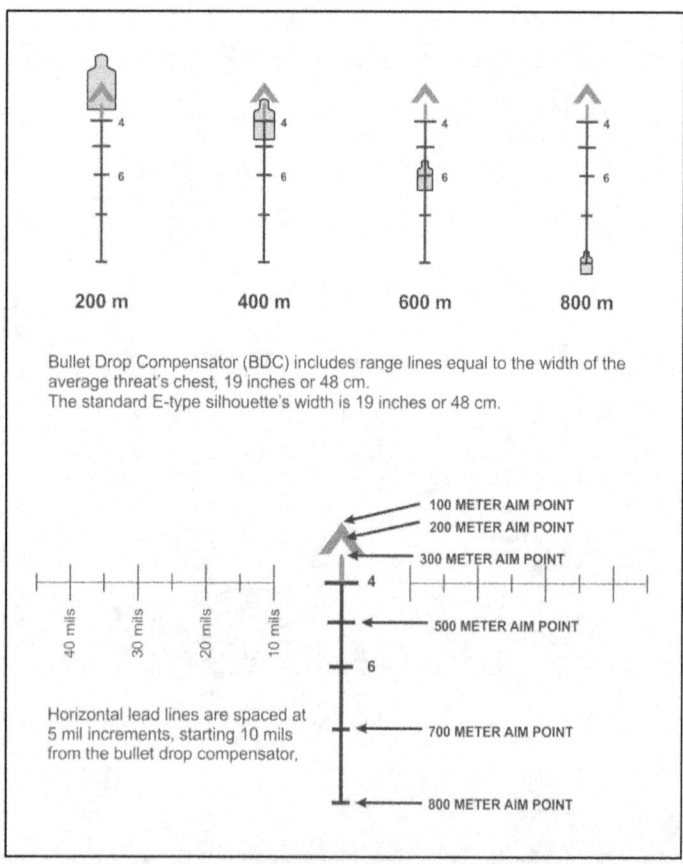

Figure C-3. RCO range determination using the bullet drop compensator reticle

Complex Engagements

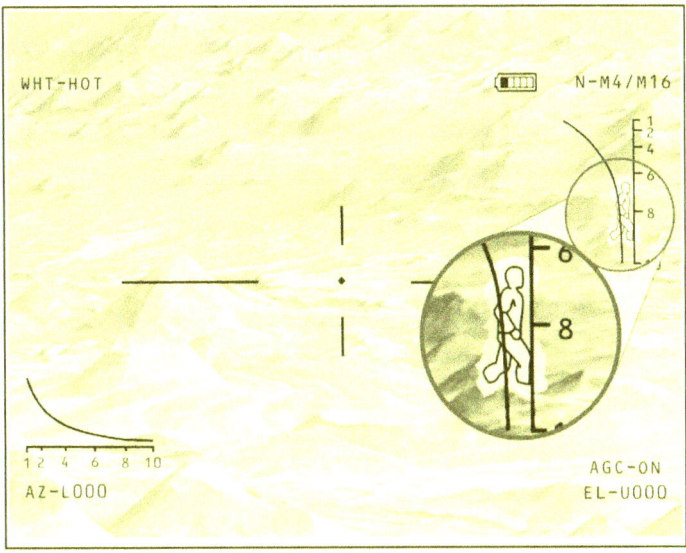

Figure C-4. Reticle relationship using a stadiametric reticle example

C-10. Anything that limits the visibility (such as weather, smoke, or darkness) will also limit the effectiveness of this method. To become proficient in using the appearance of objects method with accuracy, shooters must be familiar with the characteristic details of objects as they appear at various ranges.

Appendix C

MOVING TARGETS

C-11. Moving targets are those threats that appear to have a consistent pace and direction. Targets on any battlefield will not remain stationary for long periods of time, particularly once a firefight begins. Soldiers must have the ability to deliver lethal fires at a variety of moving target types and be comfortable and confident in the engagement techniques. There are two methods for defeating moving targets; tracking and trapping.

Tracking Method

C-12. The tracking method is used for a moving target that is progressing at a steady pace over a well-determined route. If a Soldier uses the tracking method, he tracks the target with the rifle's sight while maintaining sight alignment and a point of aim on or ahead of (leading) the target until the shot is fired.

C-13. When establishing a lead on a moving target, the rifle sights will not be centered on the target and instead will be held on a lead in front of the target. The basic lead formula for moving targets that are generally perpendicular to the shooter (moving across the sector of observation), is—

$$\frac{1}{100} R(7) = L$$

or

$$\frac{1}{100} \text{Range to Target} \times 7 = \text{Lead in Inches}$$

C-14. This formula is used to determine the baseline lead in the direction of travel of the target when its pace is approximately 3 mph or 4.5 feet per second (fps). Figure°C-5, on page C-7, shows the application of this formula at a notional moving target:

Complex Engagements

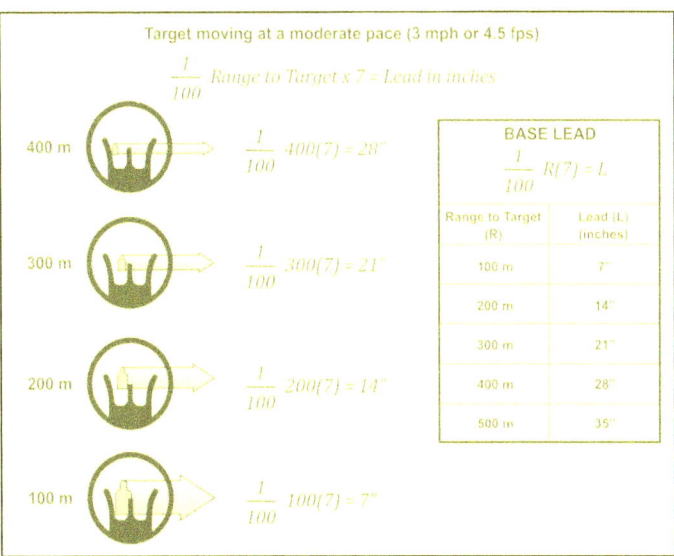

Figure C-5. Deliberate lead formula example

C-15. To execute the tracking method, a Soldier performs the following steps:
- Swing the muzzle of the rifle through the target (from the rear of the target to the front) to the desired lead (point of aim). The point of aim may be on the target or some point in front of the target depending upon the target's range, speed, and angle of movement.
- Track and maintain focus on the rifle's sight while acquiring the desired sight picture. It may be necessary to shift the focus between the rifle's sight and the target while acquiring the sight picture, but the focus must be on the rifle's sight when the shot is fired. Engage the target once the sight picture is acquired. While maintaining the proper lead,—
  - Follow-through so the lead is maintained as the bullet exits the muzzle.
  - Continue to track in case a second shot needs to be fired on the target.

Trapping Method

C-16. The trapping method (see figure C-6) is used when it is difficult to track the target with the aiming device, as in the prone or sitting position. The lead required to effectively engage the target determines the engagement point and the appropriate hold-off.

Appendix C

C-17. With the sights settled, the target moves into the predetermined engagement point and creates the desired sight picture. The trigger is pulled simultaneously with the establishment of sight picture. To execute the trapping method, a Soldier performs the following steps:

- Select an aiming point ahead of the target – where to set the trap.
- Obtain sight alignment on the aiming point.
- Hold sight alignment until the target moves into vision and the desired sight picture is established.
- Engage the target once sight picture is acquired.
- Follow-through so the rifle sights are not disturbed as the bullet exits the muzzle.

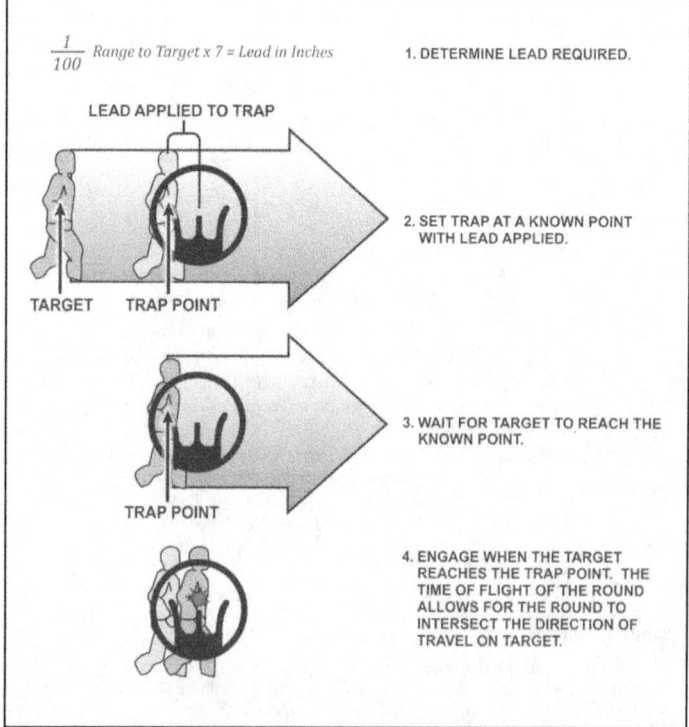

Figure C-6. Deliberate trapping method example

Complex Engagements

OBLIQUE TARGETS

C-18. Threats that are moving diagonally toward or away from the shooter are oblique targets. They offer a unique problem set to shooters where the target may be moving at a steady pace and direction; however, their oblique posture makes them appear to move slower.

C-19. Soldiers should adjust their hold-off based on the angle of the target's movement from the gun-target line. The following guide (see figure C-7) will help Soldiers determine the appropriate percentage of hold-off to apply to engage the oblique threats as they move.

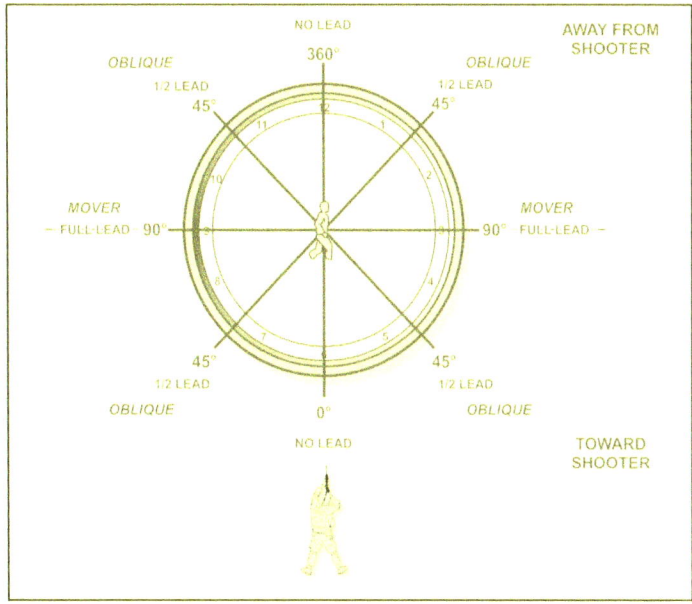

Figure C-7. Oblique target example

ENVIRONMENTAL CONDITIONS

C-20. The environment can complicate the shooter's actions during the shot process with excessive wind or requiring angled firing limited visibility conditions. Soldiers must understand the methods to offset or compensate for these firing occasions, and be prepared to apply these skills to the shot process. This includes when multiple complex conditions compound the ballistic solution during the firing occasion.

WIND

C-21. Wind deflection is the most influential element in exterior ballistics. Wind does not push the projectile causing the actual deflection. The bullet's tip is influenced in the direction of the wind slightly, resulting in a gradual drift of the bullet in the direction of the wind. The effects of wind can be compensated for by the shooter provided they understand how wind effects the projectile and the terminal point of impact. The elements of wind effects are—
- The **time** the projectile is exposed to the wind (range).
- The **direction** from which the wind is blowing.
- The **velocity** of the wind on the projectile during flight.

Wind Direction and Value

C-22. Winds from the left cause an effect on the projectile to drift to the right, and winds from the right cause an effect on the projectile to drift to the left. The amount of the effect depends on the time of (projectile's exposure) the wind speed and direction. To compensate for the wind, the firer must first determine the wind's direction and value. (See figure C-8 on page C-11.)

C-23. The clock system can be used to determine the direction and value of the wind. Picture a clock with the firer oriented downrange towards 12 o'clock.

C-24. Once the direction is determined, the value of the wind is next. The value of the wind is how much effect the wind will have on the projectile. Winds from certain directions have less effect on projectiles. The chart below shows that winds from 2 to 4°o'clock and 8 to 10 o'clock are considered full-value winds and will have the most effect on the projectile. Winds from 1, 5, 7, and 11 o'clock are considered half-value winds and will have roughly half the effect of a full-value wind. Winds from 6 and 12°o'clock are considered no-value winds and little or no effect on the projectile.

Complex Engagements

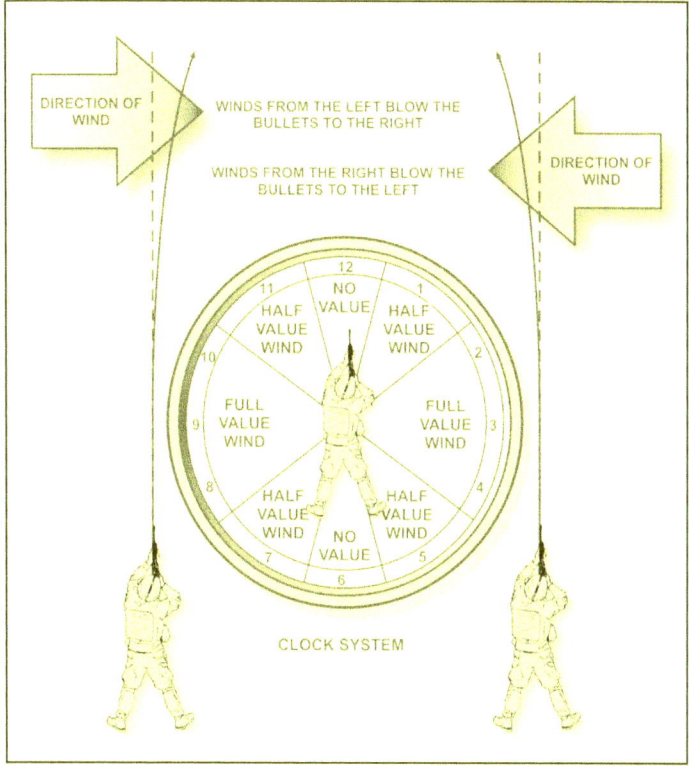

Figure C-8. Wind value

C-25. The wind will push the projectile in the direction the wind is blowing (see figure C-9). The amount of effects on the projectile will depend on the time of exposure, direction of the wind, and speed of the wind. To compensate for wind the Soldier uses a hold in the direction of the wind.

Appendix C

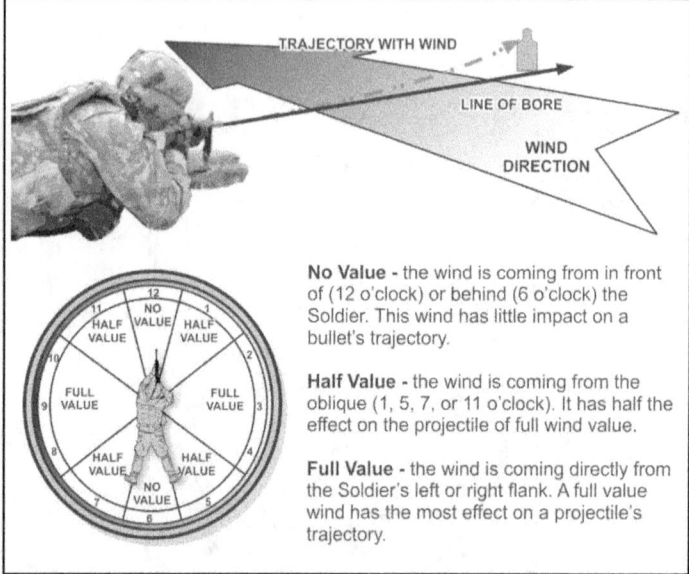

Figure C-9. Wind effects

Wind Speed

C-26. Wind speeds can vary from the firing line to the target. Wind speed can be determined by taking an average of the winds blowing on the range. The firer's focus should be on the winds between the firer and the target. The front 1/3 of the trajectory plays the most significant role in determining the bullet's wind drift deflection, but with increasing range, the firer must consider the wind speed at midpoint and the target area to make the best overall assessment.

C-27. The Soldier can observe the movement of items in the environment downrange to determine the speed. Each environment will have different vegetation that reacts differently.

C-28. Downrange wind indicators include the following:
- 0 to 3 mph = Hardly felt, but smoke drifts.
- 3 to 5 mph = Felt lightly on the face.
- 5 to 8 mph = Keeps leaves in constant movement.
- 8 to 12 mph = Raises dust and loose paper.
- 12 to 15 mph = Causes small trees to sway.

C-29. The wind blowing at the Soldiers location may not be the same as the wind blowing on the way to the target.

Wind Estimation

C-30. Soldiers must be comfortable and confident in their ability to judge the effects of the wind to consistently make accurate and precise shots. Soldiers will use wind indicators between the Soldier and the target that provide windage information to develop the proper compensation or hold-off.

C-31. To estimate the effects of the wind on the shot, Soldiers need to determine three windage factors:
- Velocity (speed).
- Direction.
- Value.

Determining Wind Drift

C-32. Once wind velocity, direction, and value have been determined, Soldiers determine how to compensate for the effects of wind. For the Soldier, there are three methods of determining the appropriate hold-off to adjust for excessive wind: using the wind formula, wind estimation, or referencing a generalized ballistic windage chart.

C-33. Once the range to target and wind speed are known, the formula below is used to determine drift. The output from the formula is in MOA. The final answer is rounded off to make the calculation quicker to perform. This formula (see figure C-10) will allow the Soldier to adjust for the distance that the wind displaces his projectile.

Appendix C

Wind Hold Estimation Formula for 5.56 mm

$$MOA\ of\ Drift = \frac{(\frac{1}{100}R)V}{7} = \frac{\frac{1}{100} Range\ (meters) \times Velocity\ (mph)}{7}$$

$$\frac{(\frac{1}{100}\ 400\ meters) \times 7\ mph}{7} = \frac{4 \times 7}{7} = 4\ MOA = 16"$$

	meters	WIND SPEED (miles per hour {mph})			
		5	10	15	20
RANGE	300	6 inches	13 inches	19 inches	26 inches
	200	3 inches	5 inches	8 inches	11 inches
	100	1 inch	1 inch	2 inches	3 inches

NOTE: This chart represents ballistic information for M855 ammunition.
MOA - minute of angle
mm - millimeter

Figure C-10. Wind formula and ballistics chart example

C-34. The ballistics chart shows the wind drift in inches at ranges from 100 meters – 300 meters and wind speeds up to 20 mph. The data from the 100-m (meter) line shows that even in a 20-mph wind there is very little deflection of the round. At 300 meters, it can be seen that the same 20-mph wind will blow the bullet 26 inches. This illustrates the fact that the bullet is effected more by the wind the further it starts out from the target.

Windage Hold

C-35. Using a hold involves changing the point of aim to compensate for the wind drift. For example, if wind causes the bullet to drift 12 inches to the left, the aiming point must be moved 12 inches to the right. (See figure C-11 on page C-15.)

Complex Engagements

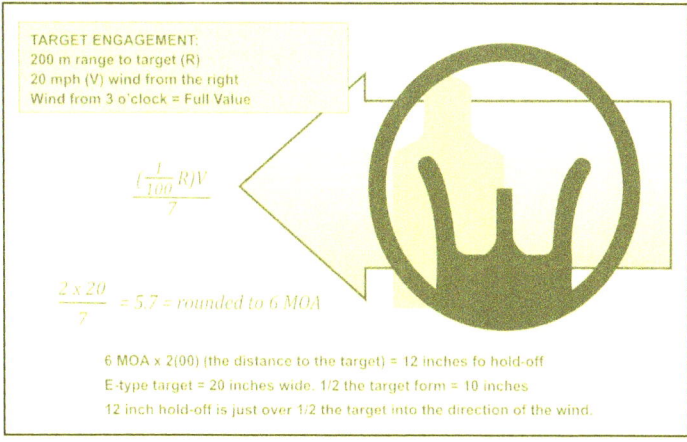

Figure C-11. Hold-off example

> *Note.* The aiming point is center mass of the visual target, allowing for the greatest possibility of impacting the target. The hold off is based on the distance *from* center mass. Soldiers apply the hold-off creating the new point of aim.

C-36. Firers must adjust their points of aim into the wind to compensate for its effects. If they miss a distant target and wind is blowing from the right, they should aim to the right for the next shot. A guide for the initial adjustment is to split the front sight post on the edge of the target facing the wind.

C-37. Newly assigned Soldiers should aim at the target's center of visible mass for the first shot, and then adjust for wind when they are confident that wind caused the miss. Experienced firers should apply the appropriate hold-off for the first shot, but should follow the basic rule—when in doubt, aim at the center of visible mass.

Angled Fire

C-38. Firing uphill or downhill at angles greater than 30 degrees, the firer must account for the change in the strike of the round from a horizontal trajectory. Rounds fired at excessive angles at extended ranges beyond the weapon's zero distance strike high on the target. To compensate for this, firers can rapidly determine a correct firing solution using the Quick High Angle Formula.

C-39. The first step is to determine the appropriate hold for the range to target beyond zero distance. Table C-1 provides the approximate holds for M855A1, 5.56mm, Ball, Enhanced Performance Round (EPR) at ranges beyond the Army standard 300 meter zero—

Table C-1. Standard holds beyond zero distance example

Range (meters)	Drop from Point of Aim (inches)	MOA Hold	Mil Hold
400	-11.9	2.6	0.7
500	-31.4	5.5	1.6
600	-59.7	8.7	2.5

C-40. Next, the firer estimates the angle of fire to either 30, 45, or 60 degrees. The firer then applies that information to the Quick High Angle Formula to determine the approximate high angle hold. This formula is built to create a rapid hold adjustment that will get the shot on target.

C-41. Figure C-12 shows the quick high angle formula with an example in both MOA and mils. The example is based on a target at 500 meters, and provides effective solutions for the three angle categories; 30, 45, and 60 degrees.

QUICK HIGH ANGLE FORMULA

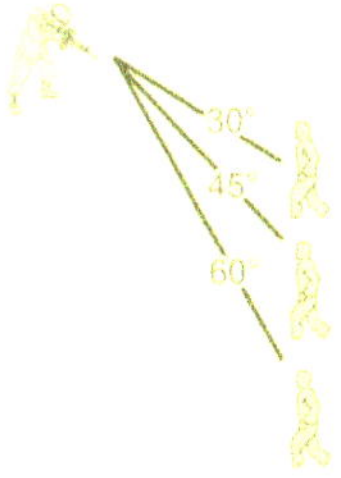

	DOWN ANGLE	MOA ADJUSTMENT/ 100 METERS	Mil ADJUSTMENT/ 100 METERS
30° 45° 60°	30°	-1/2 MOA	0.15 mils
	45°	-1 MOA	0.3 mils
	60°	-2 MOA per 100 meters, then add 1 MOA	0.6 mils per 100 meters, then add 0.3 mils

MINUTE OF ANGLE (MOA) EXAMPLE at 500 meters

DEGREES	RANGE HOLD		ANGLE OFFSET (-)		60 DEGREE OFFSET (+)		HIGH ANGLE HOLD
30°	5.5 MOA	-	2.5 MOA	+		=	3 MOA
45°	5.5 MOA	-	5 MOA	+		=	1/2 MOA
60°	5.5 MOA	-	10 MOA	+	1 MOA	=	-3 1/2 MOA

MILS EXAMPLE at 500 meters

DEGREES	RANGE HOLD		ANGLE OFFSET (-)		60 DEGREE OFFSET (+)		HIGH ANGLE HOLD
30°	1.6 mils	-	0.75 mils			=	+0.85 mils
45°	1.6 mils	-	1.5 mils			=	+0.1 mils
60°	1.6 mils	-	3 mils	+	0.3 mils	=	-1.1 mils

60° Rule of Thumb is Range Hold -2 MOA / 0.6 mil + High Angle Hold

Figure C-12. Quick high angle formula example

Appendix C

COMPOUND CONDITIONS

C-42. When combining difficult target firing occasion information, Soldiers can apply the rules specific to the situation together to determine the appropriate amount of hold-off to apply.

C-43. The example below shows the application of different moving target directions with varying speed directions. This is a general example to provide the concept of applying multiple hold-off information to determine complex ballistic solutions for an engagement. (See figure C-13.)

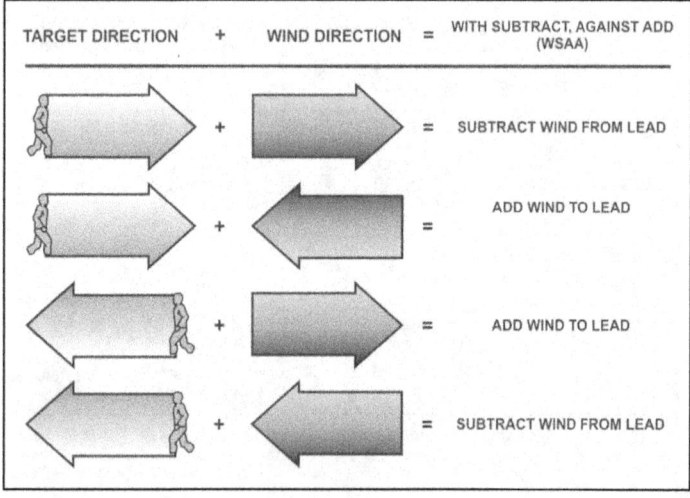

Figure C-13. Compound wind and lead determination example

Appendix D
Drills

This appendix describes the various drills for the rifle and carbine, and their purpose. The drill structure is standardized for all individual and crew served weapons in order to reinforce the most common actions all Soldiers need to routinely execute with their assigned equipment during training and combat.

These drills are used during Table III of the integrated weapons training strategy, as well as during routine maintenance, concurrent training, and during deployments. The drills found within this appendix are used to build and maintain skills needed to achieve proficiency and mastery of the weapon, and are to be ingrained into daily use with the weapon.

D-1. Each drill is designed to develop confidence in the equipment and Soldier actions during training and combat operations. As they are reinforced through repetition, they become second nature to the Soldier, providing smooth, consistent employment during normal and unusual conditions.

D-2. The drills provided are designed to build the Soldier's proficiency with the following principles:

- **Mindset** – the Soldier's ability to perform tasks quickly and effectively under stress.
- **Efficiency** – ensure the drills require the least amount of movement or steps to complete correctly. Make every step count.
- **Individual tactics** – ensure the drills are directly linked to employment in combat.
- **Flexibility** – provide drills that are not rigid in execution. Units may alter the procedural steps depending on their equipment, configuration, or tactical need.

MINDSET

D-3. Continuous combat is inherently stressful. It exhausts Soldiers and causes physiological changes that reduce their ability to perform tasks as quickly or effectively as necessary. The Soldier's ability to function under stress is the key to winning battles, since, without the Soldier, weapons and tactics are useless. Individual and unit military effectiveness depend on the Soldier's ability to think clearly, accurately, quickly, all with initiative, motivation, physical strength, and endurance.

D-4. The impact of physiological changes caused by the stress of combat escalates or de-escalates based on the degree of stimulation, causing Soldiers to attain different levels of awareness as events occur in the continually transitioning operational area around them. Maintaining a tactical mindset involves understanding one's level of awareness and transitioning between the levels of awareness as the situation requires escalation or de-escalation.

> *Note.* Stress can be countered using the principles associated with Soldier resilience and performance enhancement. The Comprehensive Soldier and Family Fitness (CSF2) is designed to increase a Soldier's ability and willingness to perform an assigned task or mission and enhance his performance by assessing and training mental resilience, physical resilience, and performance enhancement techniques and skills. This initiative introduces many resources used to train Soldiers on skills to counter stress. For more information about CSF2, see http://csf2.army.mil/.

EFFICIENCY

D-5. Efficiency is defined as the minimization of time or resources to produce a desired outcome. Efficient movements are naturally faster than movements that contain excessive or wasteful actions.

D-6. By reducing the amount of effort, mental, and/or physical, the movement becomes repeatable and the effect becomes predictable. This allows the Soldier to focus on the tactics while still maintaining the ability to produce accurate and precise fires.

INDIVIDUAL TACTICS

D-7. Individual tactics are actions independent of unit standard operating procedures (SOPs) or situations that maximize the Soldier's chance of survival and victory in a small arms, direct fire battle.

D-8. Examples of individual tactics include use of cover and standoff, or the manipulation of time and space between a Soldier and his enemy.

FLEXIBILITY

D-9. The techniques presented in this publication are not meant to be prescriptive, as multiple techniques can be used to achieve the same goal. In fact, there is no singular "one size fits all" solution to rifle fire; different types of enemies and scenarios require the use of different techniques.

D-10. However, the techniques presented are efficient and proven techniques for conducting various rifle-related tasks. Should other techniques be selected, they should meet the following criteria:

RELIABLE UNDER CONDITIONS OF STRESS

D-11. Techniques should be designed for reliability when it counts; during combat. The technique should produce the intended results without fail, under any conditions and while wearing mission-essential equipment.

Drills

D-12. It should also be tested under as high stress conditions as allowed in training.

REPEATABLE UNDER CONDITIONS OF STRESS

D-13. As combat is a stressor, a Soldier's body responds much as it does to any other stressful stimulus; physiological changes begin to occur, igniting a variable scale of controllable and uncontrollable responses based on the degree of stimulation.

D-14. The technique should support or exploit the body's natural reaction to life-threatening stress.

EFFICIENCY IN MOTION

D-15. The technique should be designed to create the greatest degree of efficiency of motion. It should contain only necessary movement. Excessive or unnecessary movement in a fighting technique costs time to execute. In a violent encounter, time can mean the difference between life and death.

D-16. Consider the speed at which violent encounters occur: An unarmed person can cover a distance of 20 feet in approximately 1 second. Efficiency decreases the time necessary to complete a task, which enhances the Soldier's safety.

DEVELOP NATURAL RESPONSES THROUGH REPETITION

D-17. When practiced correctly and in sufficient volume, the technique should build reflexive reactions that a Soldier applies in response to a set of conditions. Only with correct practice will a Soldier create the muscle memory necessary to serve him under conditions of dire stress. The goal is to create automaticity, the ability to perform an action without thinking through the steps associated with the action.

LEVERAGE OVERMATCH CAPABILITIES

D-18. Engagements can occur from 0 to 600 meters and any variance in between. Fast and efficient presentation of the rifle allows more time to stabilize the weapon, refine the aim, and control the shot required to deliver precise fires. This rapidly moves the unit toward the goal of fire superiority and gains/maintains the initiative. Speed should be developed throughout the training cycle and maintained during operations.

D-19. As distance between the Soldier and a threat decreases, so does the time to engage with well-place lethal fires. As distance increase, the Soldier gains time to refine his aim and conduct manipulations.

Appendix D

DRILLS

D-20. To build the skills necessary to master the functional elements of the shot process, certain tasks are integrated into drills. These drills are designed specifically to capture the routine, critical tasks or actions Soldiers must perform fluently and as a second nature to achieve a high level of proficiency.

D-21. Drills focus on the Soldier's ability to apply specific weapons manipulation techniques to engage a threat correctly, overcome malfunctions of the weapon or system, and execute common tasks smoothly and confidently.

DRILL A – WEAPON CHECK

D-22. The weapon check is a visual inspection of the weapon by the Soldier. A weapon check includes at a minimum verifying:
- Weapon is clear.
- Weapon serial number.
- Aiming device(s) serial number.
- Attachment points of all aiming devices, equipment, and accessories.
- Functions check.
- Proper location of all attachments on the adaptive rail system.
- Zero information.
- Serviceability of all magazines.

D-23. The weapon check is initiated when first receiving the weapon from the arms room or storage facility. This includes when recovering the weapon when they are stacked or secured at a grounded location.

D-24. Units may add tasks to Drill A as necessary. Units may direct Soldiers to execute Drill A at any time to support the unit's mission.

DRILL B – SLING/UNSLING OR DRAW/HOLSTER

D-25. This drill exercises the Soldier's ability to change the location of the weapon on demand. It reinforces their ability to maintain situational and muzzle awareness during rapid changes of the weapon's sling posture. If also provides a fitment check between the weapon, the Soldier's load bearing equipment, and the Soldier's ability to move between positions while maintaining effective use of the weapon.

D-26. When conducting this drill, Soldiers should:
- Verify the proper adjustment to the sling.
- Rotate the torso left and right to ensure the sling does not hang up on any equipment.
- Ensure the weapon does not interfere with tactical movement.

DRILL C – EQUIPMENT CHECK

D-27. This drill is a Pre-Combat Check (PCC) that ensures the Soldier's aiming devices, equipment, and accessories are prepared –

Drills

- Batteries.
- Secured correctly.
- Equipment does not interfere with tactical movement.
- Basic load of magazines are stowed properly.

DRILL D – LOAD

D-28. This is predominantly an administrative loading function. This allows the Soldier to develop reliable loading techniques.

DRILL E – CARRY (FIVE/THREE)

D-29. This is a series of five specific methods of carrying the weapon by a Soldier. These five methods are closely linked with range operations in the training environment, but are specifically tailored to combat operations. This drill demonstrates the Soldier's proficiency moving between:
- Hang.
- Safe hang.
- Collapsed low ready.
- Low ready.
- High ready (or ready up).

D-30. A leader will announce the appropriate carry term to initiate the drill. Each carry method should be executed in a random order a minimum of three times.

DRILL F – FIGHT DOWN

D-31. The Fight Down drill builds the Soldier's understanding of how to move effectively and efficiently between firing postures. This drill starts at a standing position, and, on command, the Soldier executes the next lower position or the announced position by the leader. The Fight Down drill exercises the following positions in sequence:
- Standing.
- Kneeling.
- Sitting.
- Prone.

D-32. Each position should be executed a minimum of three times. Leaders will use Drill F in conjunction with Drill G.

DRILL G – FIGHT UP

D-33. The Fight Up drill builds the Soldier's timing and speed while moving from various positions during operations. This drill starts in the prone position, and, on command, the Soldier executes the next higher position or the announced position by the leader. The Fight Up drill exercises the following positions in sequence:
- Prone.
- Sitting.

Appendix D

- Kneeling.
- Standing.

D-34. Each position should be executed a minimum of three times. Leaders will use Drill F, Fight Down, in conjunction with Drill G, Fight Up.

D-35. Leaders may increase the tempo of the drill, increasing the speed the Soldier needs to assume the next directed position. After the minimum three iterations are completed (Drill F, Drill G, Drill F, Drill G, etc.), the leader may switch between Drill F and G at any time, at varying tempo.

DRILL H – GO-TO-PRONE

D-36. The Go-To-Prone drill develops the Soldier's agility when rapidly transitioning from a standing or crouched position to a prone firing position. Standard time should be below 2 seconds.

D-37. Leaders announce the starting position for the Soldier to assume. Once the Soldier has correctly executed the start position to standard, the leader will announce GO TO PRONE. This drill should be conducted a minimum of five times stationary and five times while walking.

D-38. Leaders should not provide preparatory commands to the drill, and should direct the Soldier to go to prone when it is unexpected or at irregular intervals. Leaders may choose to include a tactical rush with the execution of Drill H.

DRILL I – RELOAD

D-39. The Tactical Reload drill is executed when the Soldier is wearing complete load bearing equipment. It provides exercises to assure fast reliable reloading through repetition at all firing positions or postures.

D-40. The Soldier should perform Drill I from each of the following positions a minimum of seven times each:
- Standing.
- Squatting.
- Kneeling.
- Prone.

D-41. Leaders may include other drills while directing Drill I to the Soldier to reinforce the training as necessary.

DRILL J – CLEAR MALFUNCTION

D-42. This drill includes the three methods to clear the most common malfunctions on a rifle or carbine in a rapid manner, while maintaining muzzle and situational awareness. Soldiers should perform all three variations of clearing a malfunction based on the commands from their leader.

Drills

D-43. Each of the three variations of Drill J should be executed five times. Once complete, leaders should incorporate Drill J with other drills to ensure the Soldier can execute the tasks at all positions fluently.

DRILL K – UNLOAD / SHOW CLEAR

D-44. This is predominantly an administrative unloading function, and allows the Soldier to develop reliable clearing techniques. This drill should be executed in tandem with Drill D, Load. It should be executed a minimum of seven times in order to rotate through the Soldier's magazine pouch capacity, and reinforce the use of a "dump pouch" or pocket, to retain expended magazines during operations.

D-45. This drill can be executed without ammunition in the weapon. Leaders may opt to use dummy ammunition or spent cartridge cases as desired. In garrison environments, Leaders should use Drill K on demand, particularly prior to entering buildings or vehicles to reinforce the Soldier's skills and attention to detail.

This page intentionally left blank.

Appendix E
Zeroing

Zeroing a weapon is not a training exercise, nor is it combat skills event. Zeroing is a maintenance procedure that is accomplished to place the weapon in operation, based on the Soldier's skill, capabilities, tactical scenario, aiming device, and ammunition. Its purpose is to achieve the desired relationship between the line of sight and the trajectory of the round at a known distance. The zeroing process ensures the Soldier, weapon, aiming device, and ammunition are performing as expected at a specific range to target with the least amount of induced errors.

For Soldiers to achieve a high level of accuracy and precision, it is critical they zero their aiming device to their weapon correctly. The Soldier must first achieve a consistent grouping of a series of shots, then align the mean point of impact of that grouping to the appropriate point of aim. Soldiers use the process described in this appendix with their weapon and equipment's technical manuals to complete the zeroing task.

BATTLESIGHT ZERO

E-1. The term battlesight zero means the combination of sight settings and trajectory that greatly reduces or eliminates the need for precise range estimation, further eliminating sight adjustment, holdover or hold-under for the most likely engagements. The battlesight zero is the default sight setting for a weapon, ammunition, and aiming device combination.

E-2. An appropriate battlesight zero allows the firer to accurately engage targets out to a set distance without an adjusted aiming point. For aiming devices that are not designed to be adjusted in combat, or do not have a bullet drop compensator, such as the M68, the selection of the appropriate battlesight zero distance is critical.

ZEROING PROCESS

E-3. A specific process should be followed when zeroing. The process is designed to be time-efficient and will produce the most accurate zero possible.

E-4. The zero process includes mechanical zero, laser borelight, 25-m grouping and zeroing, and zero confirmation out to 300 meters.

Appendix E

> *Note.* Although wind and gravity have the greatest effect on the projectile's trajectory, air density and elevation must also be taken into consideration.

LASER BORELIGHT

E-5. The borelight is an eye-safe laser that is used to boresight optics, iron sights, and aiming lasers. The borelight assists the first shot group hitting the 25-m zeroing target when zeroing the weapon. Using the borelight will save range time and require less rounds for the zeroing process. Borelighting is done with a borelight, which is centered in the bore of the weapon, and with an offset target placed 10 meters from the muzzle of the weapon.

25-M GROUPING AND ZEROING

E-6. After successfully boresighting the weapon, the next step is to perform grouping and zeroing exercises. Grouping and zeroing is done at 25 meters on a 25-m zero target or at known distance range.

25-M GROUPING

E-7. The goal of the grouping exercise is for the shooter to fire tight shot groups and consistently place those groups in the same location. Tight, consistently placed shot groups show that the firer is applying proper aiming and smooth trigger control before starting the zeroing process. The firer should not start the zeroing process until they have demonstrated their ability to group well.

25-M ZEROING

E-8. Once the firer has shown their ability to accurately group, they should begin adjusting the aiming device to move the groups to the center of the target. During the zeroing process, the firer should attempt to center their groups as much as possible. Depending on the aiming device used, there may be a zero offset that needs to be used at 25 meters. During the zeroing process it is important that the firer adjusts their groups as close to the offset mark as possible.

ZERO CONFIRMATION OUT TO 300 METERS

E-9. The most important step in the zeroing process is zero confirmation out to 300 meters. Having a 25 m zero does not guarantee a center hit at 300 meters. The only way to rely on a 300-m hit, is to confirm a 300-m zero.

E-10. Confirmation can be done on any range where Soldiers can see the impacts of their rounds. Groups should be fired and aiming devices should be adjusted. At a minimum, the confirmation should be done at 300 meters. If rounds are available, groups can be fired at various ranges to show the firers where their impact will be.

E-11. When confirming zero at ranges past 100 meters, the effects of the wind needs to be considered and acted upon, if necessary. If a zero is confirmed at 300 meters on a windy day, and then the weapon is fired at a later date in different wind conditions or no wind at all, the impact will change. (See figure E-1 on page E-3.)

Zeroing

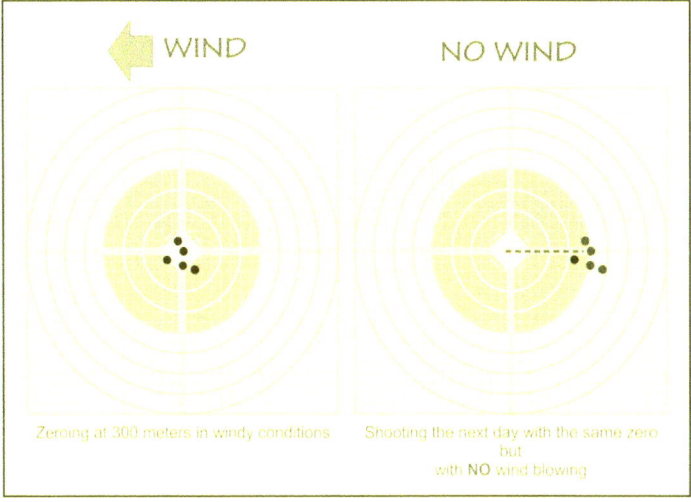

Figure E-1. Wind effects on zero at 300 meters

DOWNRANGE FEEDBACK

E-12. Feedback must be included in all live-fire training. Soldiers must have precise knowledge of a bullet strike; feedback is not adequate when bullets from previous firings cannot be identified. To provide accurate feedback, trainers ensure that Soldiers triangulate and clearly mark previous shot groups on a zeroing target or receive a hard copy from the tower on an automated range.

E-13. After zeroing, downrange feedback should be conducted. If modified field fire or known distance ranges are not available, a series of scaled silhouette targets can be used for training on the 25-m range.

E-14. With the M4- and M16-series of weapons, this range is 25 to 300 meters. This means, that with a properly zeroed rifle, the firer can aim center mass of a target between 25 meters and 300 meters and effectively engage it. A properly trained rifleman should be able to engage targets out to 600 meters in the right circumstances.

Appendix E

> *Note.* A common misconception is that wearing combat gear will cause the zero to change. Adding combat gear to the Soldier's body does not cause the sights or the reticle to move. The straight line between the center of the rear sight aperture and the tip of the front sight post either intersects with the trajectory at the desired point, or it does not. Soldiers should be aware of their own performance, to include a tendency to pull their shots in a certain direction, across various positions, and with or without combat gear. A shift in point of impact in one shooting position may not correspond to a shift in the point of impact from a different shooting position.

E-15. Figure E-2, on page E-5, shows the zeroing target for use for the M16A2/M16A4. Figure E-3, on page E-6, M4-/M16-series weapons.

Zeroing

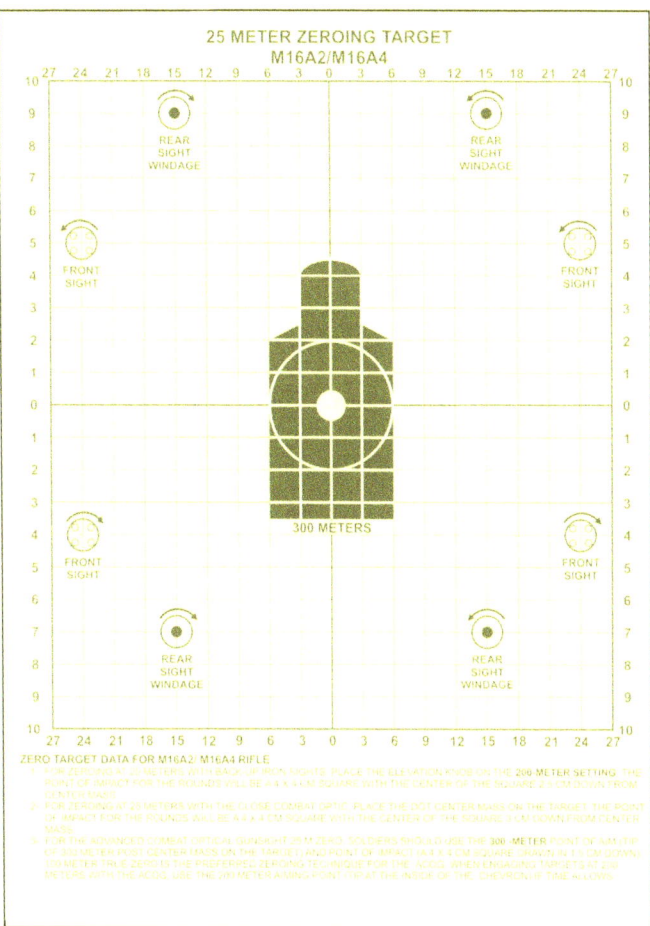

Figure E-2. M16A2 / M16A4 weapons 25m zero target

Appendix E

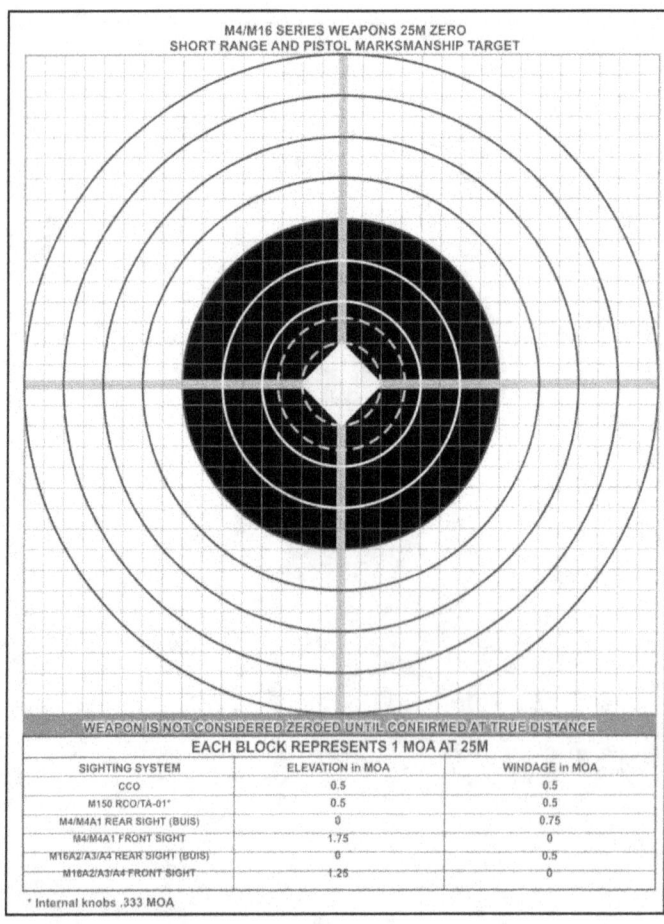

Figure E-3. M4-/M16-series weapons 25m zero short range and pistol marksmanship target

Zeroing

E-16. A good zero is necessary to be able to engage targets accurately. Whenever the Soldier deploys or does training in a new location, they should confirm the zero on their rifle if possible, as elevation, barometric pressure, and other factors will affect the trajectory of a round. There are multitudes of factors that can affect a zero, and the only sure way to know where the rounds are going, is to fire the rifle to confirm.

E-17. The zero on each assigned rifle WILL NOT transfer to another rifle. For example, if the windage zero on the Soldier's iron sights was three minutes (3MOA) left of center, putting that same setting on another rifle does not make it zeroed. This is due to the manufacturing difference between the weapons.

E-18. It is recommended that Soldiers setup their equipment and dry practice in position with gear on before coming to the range.

E-19. Standard in Training Commission (STRAC) Department of the Army Pamphlet (DA PAM) 350-38 allocates ammunition to conduct zeroing procedures using three-shot groups. The preferred method is to use a five-shot grouping, allowing the firer to more accurately analyze their shot group. Figure E-4 shows similar three-shot and five-shot groups with one shot on the right edge of the group. If all the shots were taken into account in the three-shot group, the firer would probably adjust their zero from the right edge of the four-cm circle. It is possible that the shot on the right was a poor shot and should not be counted in the group. The five-shot group on right is in the same place as the one on the left with the exception of the one shot out to the right. With four out of five shots in a tight group, the wide shot can be discounted and little or no change to the windage is necessary.

E-20. Part of the grouping and zeroing process is the marking and analysis of shot groups.

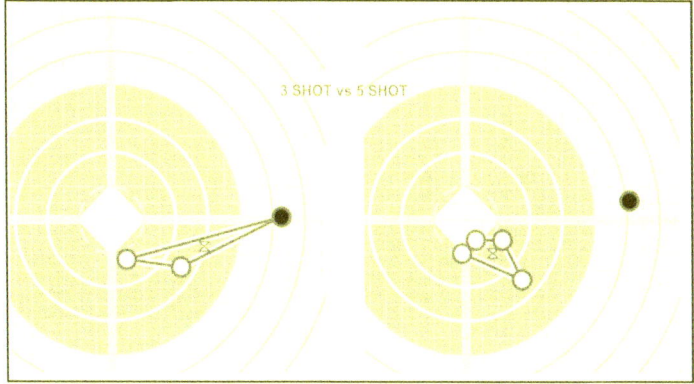

Figure E-4. Grouping

Appendix E

MARKING THE SHOT GROUP

E-21. If possible, shot groups should be marked using different colored markers so the firer can track their progress. Figure E-5 shows a technique for marking shot groups on a zero target. This technique allows the firer and coach to track their progress throughout the grouping and zeroing phase.

E-22. All sight adjustments are from the center of the group, called the ***mean point of impact (MPI)***, and not from the location of a single shot. When using five-shot group, a single shot that is outside of the rest of the group should not be counted in the group for sight adjustment purposes.

Note. This figure depicts the color variations in shades of gray.

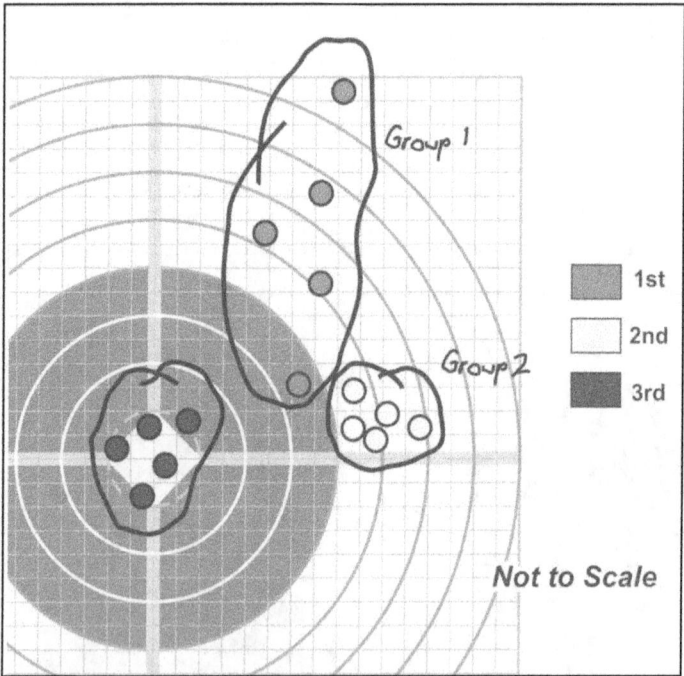

Figure E-5. Marking shot groups

Zeroing

E-23. The firer shoots and marks their first shot group with a colored marker. The color of the first group is noted by placing a line with that marker next to the 1 on the right side of the zero target. Groups are fired and marked until they are consistently in the same location.

E-24. Each sight adjustment is annotated in the same color as the group that was just fired.

COACHING

E-25. Coaching is the process of having another Soldier observe the firer during the firing process to look for shooting errors that the firer themselves may not consciously know they are making.

TYPES OF COACHES

E-26. Firing a rifle properly requires the consistent and proper application of the elements of employment. It is about doing the right thing, the same way, every shot. The small arms trainer is also the validation point for any questions during employment training. In most cases, once group training is completed, it will be the firer's responsibility to realize and correct his own firing errors but this process can be made easier through the use of a coach.

E-27. Two types of coaches exist, the experienced coach and the peer coach. Although each should execute coaching the same way, experienced coaches have a more thorough understanding of employment and should have more knowledge and practice in firing than the Soldiers they are coaching. Knowledge and skill does not necessarily come with rank therefore Soldiers serving as experienced coaches should be carefully selected for their demonstrated firing ability and their ability to convey information to firers of varying experience levels.

EXPERIENCED COACHES

E-28. Experienced coaches are generally in shorter supply throughout the Army and are generally outnumbered by less skilled firers. This lack of experienced coaches usually leads to one experienced coach watching multiple firers dependent upon the table or period of employment being fired. It often helps the experienced coach to make notes of errors they observe in shooters and discuss them after firing that group. It is often difficult for the coach to remember the errors that they observe in each and every firer.

PEER COACHES

E-29. Using a peer coach, although generally not as effective as using an experienced coach, is still a very useful technique. The advantage of using a peer coach is two-fold: a peer coach may use their limited knowledge of employment to observe the firer when an experienced coach is not available or is occupied with another firer and can either talk the firer through the shooting errors that they have observed or bring any observed shooting errors to the attention of the experienced coach. The other advantage of using a peer coach is that the peer coach themselves, through the act of coaching, may be able to observe mistakes made by the firer and learn from them before making the mistakes themselves. Many people grasp instruction more deeply when they are coaching others than when they are simply told to do something.

Note. Peer coaches can be limited by their level of training.

E-30. Except for aiming, the coach can observe most of the important aspects of the elements of employment. To determine the unobservable errors of shooting the coach and the firer must have an open dialog and there must be a relaxed environment for learning. The firer cannot be hesitant to ask questions of the coach and the coach must not become a stressor during firing. The coach must have the ability to safely move around the firer to properly observe. There is no one ideal coaching position. The following section will discuss the elements of shooting and how best to observe them as a coach.

STABILIZE

E-31. For the coach to observe how stable the shooter is, they may have to move to different sides of the shooter. To observe the shooter's non-firing elbow (to ensure it makes contact with the ground), the coach will need to be on the shooter's non-firing side. To observe the cant of the weapon (the sights on the weapon should be pointing towards 12 o'clock position, not 11 or 1 o'clock positions), the coach will need to watch the relationship of the front sight to the barrel from behind the shooter. The coach should look for all the other aspects of good positions as outlined in chapter 6 of this publication. The coach should also observe the total amount of weapon movement on recoil. A good stable position will have minimal movement under recoil.

AIMING

E-32. Determining the aspects of the firer's aiming (sight picture, sight alignment, point of focus) requires dialogue between the firer and the coach. Often, a shooter will not realize his aiming errors until he discovers them on his own. The only method a coach has to observe aiming errors is to use of an M16 sighting device (A2, left and right, DVC-T 7-84), but this device can only be used on rifles with carrying handle sights. Without the use of a sighting device, the coach must rely on drawings, discussions, or the use of an M15A1 aiming card (DVC-T 07-26) to determine where the firer is aiming on the target, his focus point during firing (which should be the front sight), and where his front sight was at the moment of firing in relation to the rear sight aperture and the point of aim on the target. The technique of having the firer call his shots should also be used. This technique involves calling the point on the target where the sights were located at the moment of firing and matching the point called with the impact locations on the target. Calling the shot helps the firer learn to focus on the front sight during the entire firing process.

E-33. When optics are being used, the shooter can tell the coach where he was holding. This is of particular importance with the RCO. Coaches must insure the 300m aim point is used when zeroing at 25-m.

Zeroing

CONTROL

E-34. The ideal position to observe trigger squeeze is from the non-firing side because the coach will have a better view of the speed of pull, finger position on the trigger, and release or pressure on the trigger after firing. The coach can look from behind the shooter to observe the barrel for lateral movement caused by slapping the trigger during firing.

COACHING FACTORS

E-35. All firing happens at the weapon. This means that the coach should be focused solely on the shooter during firing and not on what is happening down range.

E-36. There is no way for a coach to observe only the bullets impact on target and know what errors the firer made. The coach must watch the shooter during firing to determine errors and use the impacts to confirm their assumptions.

E-37. For a coach to properly observe all aspects of firing they must be able to observe the shooter, safely, from both sides and the back. There is no prescribed coaching position.

E-38. Coaching requires a relaxed atmosphere with open communication between the firer and the coach.

SHOT GROUP ANALYSIS

E-39. Shot group analysis involves the firer correlating the shots on paper with the mental image of how the shots looked when fired. An accurate analysis of the shot group cannot be made by merely looking at the holes in the paper. It is more important to observe the firer than to try and analyze the target. All firing takes place at the weapon, and the holes in the paper are only an indicator of where the barrel was pointed when the rifle was fired. When coaches are analyzing groups, they must question the firer about the group to make a determination of what caused the placement of the shots.

E-40. For example, if the firer has a tight group – minus one shot that is well outside of the group, the firer should have observed the outlying shot while firing. The firer would discount this shot when marking their group. (See figure E-6a and figure E-6b.) If a coach is analyzing the group, the firer would tell them that they performed poorly on the one shot that is out of the group.

Appendix E

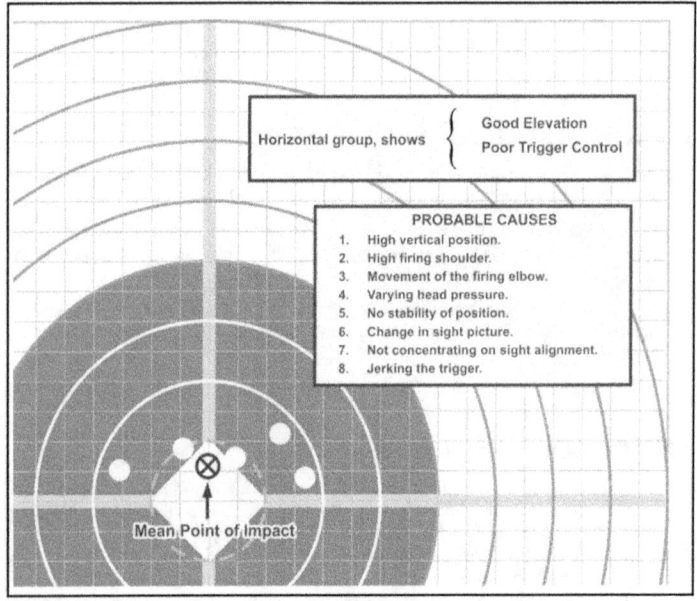

Figure E-6a. Horizontal diagnostic shots

Zeroing

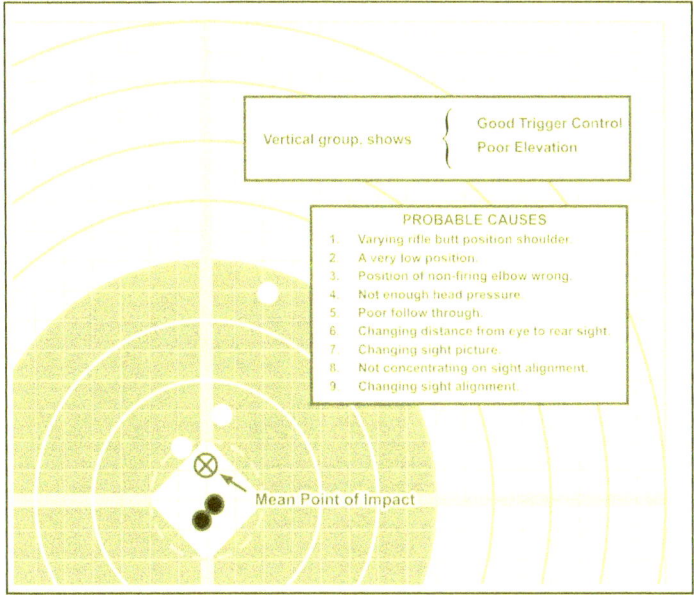

Figure E-6b. Vertical diagnostic shots

E-41. Novice shooters may benefit from not marking their own shot group. When marking a shot group an inexperienced or stressed Soldier may unintentionally make mental corrections. These mental corrections along with the mechanical corrections to their weapon will cause further issues during follow on shot groups. The experienced Soldier is less likely to make adjustments to their sight placement along with the mechanical changes to the weapon, knowing the zero process is aligning the sights to the location of the impact of the rounds. Having a coach or the employment instructor simply inform the Soldier of mechanical changes needed to the aiming device is an effective way to accomplish this method.

E-42. Observing the shooter must be accomplished before analyzing the target can become effective. Bullets strung vertically do not necessarily mean a breathing issue, nor do bullets strung horizontally absolutely indicate a trigger squeeze problem. Coaches must learn to identify shooter errors during firing and use the bullet's impacts on target to confirm their observations. There are often several firing errors that can be the cause of certain misplacements of impacts. The coach has to realize that bullets only go where

Appendix E

the barrel is pointed, so he has to determine what happened that caused the barrel to be pointed in those directions, and those causes can be many.

E-43. They key to proper coaching is becoming a shooting DETECTIVE. The coach needs to observe the shooter, question the shooter, look at the evidence down range, question the shooter again, make assumptions based upon the evidence available, and then act upon his assumptions. The coach and shooter must have a free and open dialog with each other in a relaxed atmosphere. Remember if a Soldier learns to shoot poorly they will only be capable of shooting poorly.

Displacement of Shots Within a Group (Flyers)

E-44. The capability of the weapon to shoot groups varies dependent on the number of rounds fired through the barrel over its lifetime. The average expected group size is 1 inch (approximately 2.5 centimeters) at 25 meters; some guns may shoot slightly larger than this. If a shooter is firing groups larger than a normal group size the next step should be to have a known skilled shooter attempt to fire and group with the shooter's weapon. If a proven skilled shooter is able to fire groups of the normal size it is most likely an issue with the original shooter. If however the skilled shooter cannot fire within the accepted group size there may be something wrong with the gun or barrel.

E-45. When looking at groups where there are one to two shots away from the group body (one shot away for a three round group, one or two shots away for a five round group), the coach must look objectively at the overall consistency of group placement. A bad shot or group might not indicate a poor grasp of the elements; every shooter will have a bad shot now and again, and some shooters may even have a bad group now and again. Coaches need to use their experience and determine whether or not the firer had a bad shot, a bad group, or doesn't have a clear grasp of the elements and take the necessary steps to get the shooter to the end-state. The coach may have the firer shoot again and ignore the bad group or bad shot, instead hoping that the new group matches up with the previous shot groups or the coach may need to pull the shooter off the line and cover the basic elements. Contrary to popular belief, having a firer shoot over and over again in one sitting, until the firer GETS IT RIGHT is not a highly effective technique.

Bullets Dispersed Laterally on Target

E-46. Bullets displaced in this manner could be caused by a lateral movement of the barrel due to an unnatural placement of the trigger finger on the trigger. Reasons for this could include—

- The shooter may be slightly misaligning the sights to the left and right.
- The shooter may have the sights aligned properly but may have trouble keeping the target itself perfectly centered on the tip of the front sight.
- Shooter may be closing eyes at the moment of firing or flinching.

Bullets Dispersed Vertically on Target

E-47. Bullets displaced in a vertical manner could be caused by the following:

Zeroing

- Shooter may be misaligning the front sight in the rear sight aperture vertically. May be caused by the shooter watching the target instead of the front sight. Happens more frequently from less stable positions (kneeling, unsupported positions) due to the natural movement of the weapon.
- Shooter may have trouble seeing the target and keeping the tip of the front sight exactly centered vertically on the target. Coach may consider using a larger target or a non-standard aiming point such as a 5-inch circle. Many shooters find it easier to find the center of a circle than a man shaped target.
- Shooter may not have good support, which causes him to readjust their position every shot and settle with the sights slightly misaligned.
- Shooter may be flinching or closing eyes at the moment of firing.
- Shooter may be breathing while firing the rifle. (This is not normally the case, most shooters instinctively hold their breath just before the moment of firing).

Large Groups

E-48. Large groups are most commonly caused by the shooter looking at the target instead of the front sight. This causes the shooter to place the front sight in the center of the target without regard for its location in the rear sight aperture. A small misalignment of the sights will result in a large misplacement of shots downrange.

E-49. Most likely it is not a point of aim issue; most shooters will not fire when their properly aligned sights are pointed all over the target.

Good Groups That Change Position on the Target

E-50. When the shooter has good groups but they are located at different positions on the target, there can be a number of reasons. These include the following:
- May be caused by the shooter properly aligning sights during shooting but picking up a different point of aim on the target each time.
- May be caused by the shooter settling into a position with the front sight on target but the sights misaligned. The shooter maintains the incorrect sight picture throughout the group but aligns the sights incorrectly and in a different manner during the next group. Tell the firer to focus on the front sight and have them check natural point of aim before each group.

This page intentionally left blank.

Glossary

The glossary lists acronyms and terms with Army or joint definitions. Where Army and joint definitions differ, (Army) precedes the definition. Terms for which TC 3-22.9 is the proponent are marked with an asterisk. The proponent manual for other terms is listed in parentheses after the definition.

SECTION I – ACRONYM/ABBREVIATIONS

AM	arc of movemet
ARNG	Army National Guard
ARNGUS	Army National Guard of the United States
ARS	adapter rail system
ATPIAL	advanced target pointer illuminator aiming light
BC	ballistic coefficient
BDC	bullet drop compensater
BUIS	back up iron sight
BZO	battle sight zero
CBRN	chemical, biological, radiological, and nuclear
CCO	close combat optic
CSF2	Comprehensive Soldier and Family Fitness
CoVM	center of visible mass
DA	Department of the Army
DBAL-A2	dual beam aiming laser-advanced2
DMC	digital magnetic compass
DOTD	Directorate of Training and Doctrine
DODIC	Department of Defense Identification Code
EENT	end evening nautical twilight
Ek	kinectic energy
fps	feet per second
FOV	field of view
GTL	gun target line
HTWS	heavy thermal weapons sight
I2	image intensifier
IR	infrared
LASER	light amplified stimulated emitted radiation
LCD	liquid crystal display
LRF	laser range finder
LWTS	light weapons thermal sight
MASS	modular accessory shotgun system

Glossary

MCoE	United States Army Maneuver Center of Excellence
METT-TC	mission, enemy, terrain and weather, troops and support-time available, and civil considerations
MIL STD	military standard
m	meter
mm	millimeter
mph	mile per hour
MOA	minutes of angle
MTBF	mean time between failures
MWO	modified word order
MWS	modular weapon system
MWTS	medium weapon thermal sight
NATO	North Atlantic Treaty Organization
NOD	night observation device
PAM	pamphlet
PMCS	preventative maintenance checks and services
POA	point of aim
POI	point of impact
NSN	National Stock Number
RCO	rifle combat optic
SAA	small arms ammunition
SOP	standard operating procedure
STANAG	Standardized Agreement
STRAC	Standard in Training Commission
STORM	illuminator, integrated, small arms
TACSOP	tactical standard operating procedure
TC	Training Circular
TES	tactical engagement simulation
TM	Technical Manual
T	time
TWS	thermal weapon sight
µm	micrometer
USAR	United States Army Reserve
U.S.	United States
VAL	visible aim laser
VFG	vertical foregrip
V	velocity
WCS	weapon control status
WTS	weapons thermal sights

SECTION II – TERMS

drag

The friction that slows the projectile down while moving through the air.

employment

The application of the functional elements of the shot process and skills to accurately and precisely fire a weapon at stationary or moving targets.

shot process

The basic outline of an engagement sequence all firers consider during any engagement.

stability element

The physical relationship between the weapon system, the shooter's body, the ground, and any other objects touching the weapon or shooter's body.

trajectory

The flight path that the bullet takes from the rifle to the target.

This page intentionally left blank.

References

REQUIRED PUBLICATIONS

ADRP 1-02, *Terms and Military Symbols*, 7 December 2015.
JP 1-02, *Department of Defense Dictionary of Military and Associated Terms*, 8 November 2010.

RELATED PUBLICATIONS

Most Army doctrinal publications and regulations are available at: http://www.apd.army.mil.

Most joint publications are available online at: http://www.dtic.mil/doctrine/doctrine/doctrine.htm.

Other publications are available on the Central Army Registry on the Army Training Network, https://atiam.train.army.mil.

Military Standards are available online at http://quicksearch.dla.mil.

- ATP 3-21.8, The Infantry Rifle Platoon and Squad of the Infantry Brigade Combat Team, 12 April 2016.
- DA PAM 350-38, *Standards in Training Commission*, 6 October 2015.
- FM 27-10, *The Law of Warfare*, 18 July 1956.
- MIL-STD-709D, *Department of Defense Criteria Standard: Ammunition Color Coding*, 16 March 2009.
- MIL-STD-1913, *Military Standard Dimensioning of Accessory Mounting Rail for Small Arms Weapons*, 3 February 1995.
- TC 3-22.12, *M26 Modular Accessory Shotgun System*, 12 November 2014.
- TM 9-1005-319-10, *Operator's Manual for Rifle, 5.56 MM, M16A2 W/E (NSN 1005-01-128-9936) (EIC: 4GM); Rifle, 5.56 MM, M16A3 (1005-01-357-5112); Rifle, 5.56 MM, M16A4 (1005-01-383-2872) (EIC: 4F9); Carbine, 5.56 MM, M4 W/E (1005-01-231-0973) (EIC: 4FJ); Carbine, 5.56 MM, M4A1 (1005-01-382-0953) (EIC: 4GC) {TO 11W3-5-5-41; SW 370-BU-OPI-010}*, 30 June 2010.
- TM 9-1240-413-13&P, *Operator and Field Maintenance Manual Including Repair Parts and Special Tool List for M68 Sight, Reflex, w/Quick Release Mount and Sight Mount Close Combat Optic (CCO) (NSN 1240-01-411-1265) (NSN 1240-01-540-3690) (NSN 1240-01-576-6134) {AF TO 11W3-5-5-121}*, 4 May 2013.
- TM 9-1240-416-13&P, *Operator and Field Maintenance Manual Including Repair Parts and Special Tool List for the M150 Sight, Rifle Combat Optic (RCO) (NSN: 1240-01-557-1897)*, 21 June 2013.
- TM 9-1300-200, *Ammunition, General*, 3 October 1969.

References

TM 9-1305-201-20&P, *Unit Maintenance Manual (Including Repair Parts and Special Tools List) for Small Arms Ammunition to 30 Millimeter Inclusive (Federal Supply Class 1305)*, 5 October 1981.

TM 9-5855-1912-13&P, *Operator and Field Maintenance Manual Including Repair Parts and Special Tools List for Dual Beam Aiming Laser-Advanced2 (DBAL-A2), AN/PEQ-15A (NSN: 5855-01-535-6166) (NSN: 5855-01-579-0062) (LIN: J03261)*, 1 September 2012.

TM 9-5855-1913-13&P, *Operator and Field Maintenance Manual Including Repair Parts and Special Tools List for the Illuminator, Integrated, Small Arms (STORM) AN/PSQ-23 TAN (NSN: 5855-577-5946 (4XG) (NSN: 5855-01-535-1905) (EIC: 4XF) (LIN: J68653)*, 31 August 2012.

TM 9-5855-1914-13&P, *Operator and Field Maintenance Manual Including Repair Parts and Special Tools List for the Advanced Target Pointer Illuminator Aiming Light (ATPIAL) AN/PEQ-15 (NSN 5855-01-534-5931) (NSN 5855-01-577-7174) {TM 10470B-0I/1}*, 10 September 2012.

TM 9-5855-1915-13&P, *Operator and Field Maintenance Manual (Including Repair Parts and Special Tools List) for the Target Pointer Illuminator/Aiming Light (TPIAL) AN/PEQ-2A (NSN: 5855-01-447-8992) (EIC: N/A) AN/PEQ-2B (5855-01-515-6904) (EIC: N/A) {TM 10470A-0I/1}*, 31 August 2007.

TM 11-5855-306-10, *Operator Manual for Monocular Night Vision Device (MNVD) AN/PVS-14 (NSN 5855-01-432-0524) (EIC: IPX) {TO 12S10-2PVS14-1; TM 10271A-OR/1B}*, 1 October 2010.

TM 11-5855-312-10, *Operator's Manual for Sight, Thermal AN/PAS-13B(V)2 (NSN 5855-01-464-3152) (EIC:N/A); AN/PAS-13B(V)3 (5855-01-464-3151) (EIC:N/A) {TM 10091B/10092B-10/1}*, 15 February 2005.

TM 11-5855-316-10, *Operator's Manual AN/PAS-13C(V)1 Sight, Thermal (NSN 5855-01-523-7707) (EIC: N/A) AN/PAS-13C(V)2 Sight, Thermal (NSN 5855-01-523-7713) (EIC: N/A) AN/PAS-13C(V)3 Sight, Thermal (NSN 5855-01-523-7715) (EIC: N/A)*, 31 August 2010.

TM 11-5855-317-10, *Operator's Manual for Sight, Thermal AN/PAS-13D(V)2 (NSN 5855-01-524-4313) (EIC: JH5) (MWTS) AN/PAS-13D(V)3 (NSN 5855-01-524-4314) {TM 10091C/10092C-OR/1}*, 15 May 2009.

TM 11-5855-324-10, *Operator's Manual for Sight, Thermal AN/PAS-13D(V)1 (NSN 5855-01-524-4308) (EIC: JG8) (LWTS)*, 15 May 2009.

PRESCRIBED FORMS

This section contains no entries.

REFERENCED FORMS
Unless otherwise indicated, DA forms are available on the Army Publishing Directorate (APD) web site (http://www.apd.army.mil).

DA Form 2028, *Recommended Changes to Publications and Blank Forms*.

This page intentionally left blank.

Index

Entries are listed by paragraph numbers unless mentioned otherwise.

C

carry positions

  six, 6-12

coaching

  elements of shooting, E-30

  aiming, E-32

  control, E-34

  stabilize, E-31

  two types, E-27

F

firing positions

  twelve, 6-34

M

movement techniques

  horizontal

    eight, 9-3

S

small arms ammunition, A-14

T

targets

  three threat levels, 5-7

thermal weapon sight

  five functional groups, 3-42

TC 3-22.9
13 May 2016

By Order of the Secretary of the Army:

MARK A. MILLEY
*General, United States
Army Chief of Staff*

Official:

GERALD B. O'KEEFE
*Administrative Assistant to the
Secretary of the Army*
1612002

DISTRIBUTION: *Active Army, Army National Guard, and U.S. Army Reserve:* To be distributed in accordance with the initial distribution number (IDN) 110187, requirements for TC 3-22.9.

This page intentionally left blank.

PIN: 106419-000

www.ingramcontent.com/pod-product-compliance
Lightning Source LLC
Chambersburg PA
CBHW070314190526
45169CB00005B/1629